Reducing Technological Risk Through Diligence

James R. Phimister, Vicki M. Bier, Howard C. Kunreuther, Editors

NATIONAL ACADEMY OF ENGINEERING
OF THE NATIONAL ACADEMIES

THE NATIONAL ACADEMIES PRESS
Washington, D.C.
www.nap.edu

THE NATIONAL ACADEMIES PRESS • 500 Fifth Street, N.W. • Washington, D.C. 20001

Funding for the activity that led to this publication was provided by: National Science Foundation, Public Entity Risk Institute, U.S. Nuclear Regulatory Commission, and National Aeronautics and Space Administration.

International Standard Book Number 0-309-09216-7 (Book)
International Standard Book Number 0-309-53218-3 (PDF)

Library of Congress Catalog Card Number 2004110743

Copies of this report are available from National Academies Press, 2101 Constitution Avenue, N.W., Lockbox 285, Washington, D.C. 20055; (800) 624-6242 or (202) 334-3313; Internet, *http://www.nap.edu.*

Printed in the United States of America

THE NATIONAL ACADEMIES

Advisers to the Nation on Science, Engineering, and Medicine

The **National Academy of Sciences** is a private, nonprofit, self-perpetuating society of distinguished scholars engaged in scientific and engineering research, dedicated to the furtherance of science and technology and to their use for the general welfare. Upon the authority of the charter granted to it by the Congress in 1863, the Academy has a mandate that requires it to advise the federal government on scientific and technical matters. Dr. Bruce M. Alberts is president of the National Academy of Sciences.

The **National Academy of Engineering** was established in 1964, under the charter of the National Academy of Sciences, as a parallel organization of outstanding engineers. It is autonomous in its administration and in the selection of its members, sharing with the National Academy of Sciences the responsibility for advising the federal government. The National Academy of Engineering also sponsors engineering programs aimed at meeting national needs, encourages education and research, and recognizes the superior achievements of engineers. Dr. Wm. A. Wulf is president of the National Academy of Engineering.

The **Institute of Medicine** was established in 1970 by the National Academy of Sciences to secure the services of eminent members of appropriate professions in the examination of policy matters pertaining to the health of the public. The Institute acts under the responsibility given to the National Academy of Sciences by its congressional charter to be an adviser to the federal government and, upon its own initiative, to identify issues of medical care, research, and education. Dr. Harvey V. Fineberg is president of the Institute of Medicine.

The **National Research Council** was organized by the National Academy of Sciences in 1916 to associate the broad community of science and technology with the Academy's purposes of furthering knowledge and advising the federal government. Functioning in accordance with general policies determined by the Academy, the Council has become the principal operating agency of both the National Academy of Sciences and the National Academy of Engineering in providing services to the government, the public, and the scientific and engineering communities. The Council is administered jointly by both Academies and the Institute of Medicine. Dr. Bruce M. Alberts and Dr. Wm. A. Wulf are chairman and vice chairman, respectively, of the National Research Council.

www.national-academies.org

COMMITTEE ON ACCIDENT PRECURSORS

VICKI BIER, *co-chair,* University of Wisconsin, Madison
HOWARD KUNREUTHER, *co-chair*, University of Pennsylvania, Philadelphia
JOHN F. AHEARNE, Sigma Xi, Research Triangle Park, North Carolina
ROBERT FRANCIS, Zucker, Scoutt and Rasenberger, Washington, D.C.
HAROLD S. KAPLAN, Columbia Presbyterian Medical Center, New York, New York
HENRY MCDONALD, University of Tennessee, Chattanooga
ELIZABETH MILES, Johnson and Johnson, New Brunswick, New Jersey
ELISABETH PATÉ-CORNELL, Stanford University, Stanford, California

NAE Staff

JAMES R. PHIMISTER, J. Herbert Hollomon Fellow, Program Office
PROCTOR REID, Associate Director, Program Office
NATHAN KAHL, Senior Project Assistant, Program Office
CAROL R. ARENBERG, Managing Editor

Preface

Almost every year there is at least one technological disaster that highlights the challenge of managing technological risk. On February 1, 2003, the space shuttle *Columbia* and her crew were lost during reentry into the atmosphere. In the summer of 2003, there was a blackout that left millions of people in the northeast United States without electricity. Forensic analyses, congressional hearings, investigations by scientific boards and panels, and journalistic and academic research have yielded a wealth of information about the events that led up to each disaster, and questions have arisen. Why were the events that led to the accident not recognized as harbingers? Why were risk-reducing steps not taken?

This line of questioning is based on the assumption that signals before an accident can and should be recognized. To examine the validity of this assumption, the National Academy of Engineering (NAE) undertook the Accident Precursors Project in February 2003. The project was overseen by a committee of experts from the safety and risk-sciences communities. Rather than examining a single accident or incident, the committee decided to investigate how different organizations anticipate and assess the likelihood of accidents from accident precursors.

The project culminated in a workshop held in Washington, D.C., in July 2003. This report includes the papers presented at the workshop, as well as findings and recommendations based on the workshop results and committee discussions. The papers describe precursor strategies in aviation, the chemical industry, health care, nuclear power and security operations. In addition to current practices, they also address some areas for future research.

Using accident precursors to predict and prevent accidents is not a new idea. Two industry programs, the Accident Sequence Precursors Program overseen by the U.S. Nuclear Regulatory Commission and the Aviation Safety Reporting System operated by the National Aeronautics and Space Administration, have been in existence for several decades; many other industry-specific programs have been established since those programs were started. Research has also been under way for several decades, some of which was highlighted in two earlier workshop publications, *Near-Miss Reporting as a Safety Tool* (Van der Schaaf et al., 1991) and *Accident Precursors and Probabilistic Risk Assessment* (Bier, 1998). Indeed, research results have increasingly been incorporated into practice. Nevertheless, a reassessment and affirmation of the objectives, challenges, limitations, and opportunities of precursor strategies is in order. This report is intended to address that need.

STATEMENT OF TASK

The committee was asked to meet two objectives: (1) to provided a common lexicon and framework for precursors with cross-industry applicability; and (2) to document and highlight the success of systems that have benefited from precursor information.

The committee attempted to facilitate cross-industry communication and suggest tools and terminology (see Appendixes A and D) to encourage dialogue but did not espouse a particular definition of precursors, a particular framework, or a specific approach. The keynote speakers provided an overview of the issues of precursor management (Section II); subsequent speakers discussed how different approaches could be used for risk assessment (Section III), risk management (Section IV), and for linking risk assessment and risk management on an organizational or national level (Section V).

REFERENCES

Bier, V.M., ed. 1998. Accident Sequence Precursors and Probabilistic Risk Assessment. Madison, Wis.: University of Wisconsin Press.

Van der Schaaf, T.W., D.A. Lucas, and A.R. Hale. 1991. Near-Miss Reporting as a Safety Tool. London: Butterworth-Heinemann.

Acknowledgments

This report reflects the contributions of many individuals to whom the editors extend their gratitude. First, we wish to thank the members of the committee whose involvement, participation, and interest ensured the timely and successful completion of the project. Second, we thank the workshop speakers, whose contributions cannot be overstated. The quality of their papers is a testament to their efforts. We are also grateful to the sponsors of this project, the National Science Foundation, Public Entity Risk Institute, and U.S. Nuclear Regulatory Commission. They provided support for project staff, administrative and editorial services, committee meetings, the workshop and follow-up meetings, and the compilation and publication of this report. Finally, we extend our thanks to the NAE staff who helped in the day-to-day operations of the project and in assembling and publishing the report. Proctor Reid ensured that the project was conducted in accordance with NAE policies. Nathan Kahl coordinated the logistics of the workshop. Carol Arenberg edited the manuscript and oversaw publication of the report. Penny Gibbs and Vivienne Chin provided administrative support throughout the project.

James Phimister
Project Director

Vicki Bier
Committee Co-chair

Howard Kunreuther
Committee Co-chair

Review Process

This report has been reviewed by individuals chosen for their diverse perspectives and technical expertise, in accordance with procedures approved by the National Research Council Report Review Committee. The purpose of this independent review was to provide candid, critical comments to assist NAE in making the published report as sound as possible and to ensure that the report meets institutional standards of objectivity, evidence, and responsiveness to the study charge. The review comments and the draft manuscript remain confidential to protect the integrity of the deliberative process.

The review of this report was overseen by NAE member Harold Forsen, appointed by the NAE Executive Office. Dr. Forsen was responsible for ensuring that the independent review was carried out in accordance with institutional procedures and that all review comments were carefully considered. The reviewers were: Robert Coovert, Nuclear Operations, Exelon Generation Company; Elisabeth Drake, Professor Emeritus, Massachusetts Institute of Technology; William R. Freudenburg, Professor of Environmental Studies, University of California at Santa Barbara; Deborah Grubbe, Corporate Director, Safety and Health, DuPont Company; Granger Morgan, Lord Chair Professor in Engineering, Professor and Department Head, Engineering and Public Policy, Carnegie Mellon University; D. Warner North, President, NorthWorks, Inc.; and Edwin Zebroski, Independent Consultant.

Responsibility for the final content of the report rests entirely with the authoring committee and the institution.

Contents

SECTION IV: RISK MANAGEMENT

SECTION V: LINKING RISK ASSESSMENT AND RISK MANAGEMENT

APPENDIXES

Section I
Committee Summary Report

The Accident Precursors Project

Overview and Recommendations

In 2003, the National Academy of Engineering Program Office undertook the Accident Precursors Project to examine the complex issue of accident precursor analysis and management. This seven-month project was designed to document and promote industrial and academic approaches to detecting, analyzing, and benefiting from accident precursors, as well as to understand public-sector and private-sector roles in using precursor information. The committee examined an array of approaches for benefiting from precursor information and discussed these approaches in a workshop held on July 17 and 18, 2003, in Washington, D.C. This report is the official record of the project and the workshop.

THE ACCIDENT PRECURSORS WORKSHOP

The workshop brought together experts on risk, engineers, practitioners, and policy makers from the aerospace, aviation, chemical, health care, and nuclear industries. Participants were selected for their expertise and their interest in engaging in a cross-industry dialogue. Presentations by invited experts in the field were followed by targeted discussions in breakout groups.

The workshop presentations addressed four general areas:

- The Opportunity of Precursor Analysis (Section II): the opportunities presented by precursors and some organizational and analytical approaches to detecting and analyzing them

- Risk Assessment (Section III): methods of identifying and modeling different types of precursors
- Risk Management (Section IV): how risks can be understood and mitigated once precursors have been identified and how organizations can engage their members in this endeavor
- Linking Risk Assessment and Risk Management (Section V): how linking risk assessment and risk management can create a continuous improvement process and how industry and government can facilitate this

Breakout and plenary sessions involved discussions by participants focused on advising both private organizations and government agencies on how they might use precursor information to reduce their risk exposure. Discussions were based on drafts of presenters' papers (provided before the workshop) and were led by facilitators and designated presenters.

The Committee on Accident Precursors evaluated the presentations and discussions, as well as additional submissions from Drs. Frosch and Westrum (Appendixes A and B). The resulting findings and recommendations are based on these inputs and subsequent committee deliberations.

Keynote Addresses

James Bagian, director of the Department of Veterans Affairs (VA) National Center for Patient Safety, delivered the opening keynote address. Bagian drew on his personal experiences as well as efforts by the VA to promote patient safety. He described the challenges to engaging individuals and organizations, the difficulty of recognizing when current safety standards are inadequate, and the importance of making commitments to the institutional and management processes necessary to achieving lasting, continuous improvements in safety.

Elisabeth Paté-Cornell, chair of the Department of Management Science and Engineering at Stanford University, delivered the dinner keynote address. Paté-Cornell highlighted past examples of the management of precursors. In some cases, precursors were ignored, and catastrophes followed. In other cases, precursors were recognized as warning signs, and disasters may have been avoided. Paté-Cornell also provided an overview of some of the precursor models she and her students have developed for use as decision aids. These models have been used in a broad range of applications, from optimizing the alert thresholds of warning systems, such as fire alarms (Paté-Cornell, 1986), to aiding in combating terrorism (Paté-Cornell and Guikema, 2002).

Presentations

Workshop presenters discussed how precursors could be identified and managed. Michal Tamuz of the University of Tennessee Health Science Center discussed similarities and differences in approaches to collecting and assessing precursor data in the aviation, health care, and nuclear power industries, among others. William Corcoran, president of the Nuclear Safety Review Concepts Corporation, used historical examples to illustrate the distinctions between different kinds of precursors. Martin Sattison, manager of the Risk, Reliability and Regulatory Support Department at the Idaho National Engineering and Environmental Laboratory, provided a historical overview of the U.S. Nuclear Regulatory Commission (U.S. NRC) Accident Sequence Precursor (ASP) Program and outlined lessons that could be transferred to other industries.

The next group of speakers described organizational barriers to, and opportunities for, leveraging precursor information to reduce the likelihood of accidents. Dennis Hendershot, senior technical fellow of the Rohm and Haas Company, provided everyday and industrial examples illustrating how systems can be designed or redesigned to make them inherently safer. Tjerk van der Schaaf of the Eindhoven University of Technology pointed out potential "blind spots" in reporting systems, showing why many types of near misses can go unreported. John Carroll of the Sloan School of Management of the Massachusetts Institute of Technology discussed how knowledge about potential accidents could be shared throughout an organization, both formally and informally.

The last group of speakers described approaches to engaging stakeholders, institutions, and industries in the process of identifying and managing accident precursors. Linda Connell, director of the National Aeronautics and Space Administration (NASA) Aviation Safety Reporting System (ASRS), described the history and implementation of ASRS and discussed its potential applicability in the health care, nuclear power, maritime, and security domains. Christopher Hart, assistant administrator for the Federal Aviation Administration (FAA) Office of System Safety, identified the hurdles to improving an already high level of safety (a "plateau") and discussed how a recognition of precursors could help to achieve this end. Yacov Haimes, director of the Center for Risk Management of Engineering Systems of the University of Virginia, discussed the transferability of methods used to identify and mitigate accident precursors to security systems for combating terrorism.

INTRODUCTION

In the aftermath of catastrophes, it is common to find prior indicators, missed signals, and dismissed alerts that, had they been recognized and appropriately managed before the event, might have averted the undesired event. Indeed, the accident literature is replete with examples, including the space shuttle *Columbia* (CAIB, 2003), the space shuttle *Challenger* (Vaughan, 1997), Three Mile Island (Chiles, 2002), the Concorde crash (BEA, 2002), the London Paddington train crash (Cullen, 2000), and American Airlines Flight 587 to Santo Domingo

(*USA Today*, May 25, 2003), among many others (Kletz, 1994; Marcus and Nichols, 1999; Turner and Pidgeon, 1997).

Recognizing signals *before* an accident clearly offers the potential of improving safety, and many organizations have attempted to develop programs to identify and benefit from *accident precursors*. In this summary, the committee examines how these programs can be designed to reduce system risk exposure and the responsibilities of various constituents in implementing or facilitating these programs.

At first glance it might appear that the design and operation of precursor programs would be relatively straightforward. This perception may be the result of *hindsight bias* (Fischhoff, 1975; Hawkins and Hastie, 1990), that is, after an accident, individuals often believe that the accident should have been considered highly likely, if not inevitable, by those who observed the system prior to the accident. (Hindsight bias also helps to explain the frequent discrepancies between pre- and post-accident risk assessments.)

In fact, upon examination, designing and running a precursor management program turns out to be challenging. In order to leverage precursor information, precursor programs must be able to identify possible threats before they occur; detect, filter, and prioritize precursors when they occur; evaluate precursor causes; and identify and implement corrective actions (see for example Lakats and Paté-Cornell, in press).

Although creating programs with all of these features can be difficult, it is important to consider how it can be done and whether existing programs can be improved. For example, are some individuals, companies, organizations, or even industries better able to envision and respond to potential accidents than others? If so, what processes do they use, and what organizational structures, management approaches, and regulatory frameworks support these processes?

The first topic addressed in this summary is the opportunity presented by accident precursors for improving safety. Next, a case is made, based on historical examples, for the need for a better understanding of precursor management. This is followed by several examples of precursor programs illustrating differences and parallels in approaches. The final section includes the committee's findings and recommendations.

Defining Accident Precursors

Accident precursors can be defined in a number of ways. To encourage a wide-ranging discussion of alternative definitions and reporting systems, the committee deliberately chose a broad definition. Precursors were defined as the conditions, events, and sequences that precede and lead up to accidents. Based on this definition, precursor events can be thought of loosely as "building blocks" of accidents and can include both events internal to an organization (such as equipment failures and human errors) and external events (such as earthquakes and hurricanes). This definition helped the committee (and the workshop participants) focus their discussions on the management of events that could progress to accidents, without unduly limiting or foreclosing those discussions. The definition also helped the committee and workshop participants distinguish between actual events and general underlying conditions (such as an organization's culture) that may not be part of a specific accident scenario but may still influence the likelihood of an accident.

Some organizations, such as the U.S. NRC, have chosen to limit the use of the term "precursors" to events that exceed a specified level of severity. For example, precursors might be defined as the complete failure of one or more safety systems and/or the partial failure of two or more safety systems. Similarly, a quantitative threshold may be established for the conditional probability of an accident given a precursor, and events of lesser severity either not considered precursors, or at least not singled out as deserving of further analysis.

Other organizations have designed and implemented incident reporting systems that address incidents with a much wider range of severities, including defects or off-normal events that may involve inconsequential losses of safety margins. In such cases, of course, screening, filtering, and prioritizing reported incidents is necessary to identify the events that merit further analysis; in addition, there must be a recognition that the reporting of an event is not necessarily a prejudgment of its risk significance.

Both approaches to defining precursors have advantages and disadvantages. Setting the threshold for reporting too high or defining reportable precursors too precisely may mean that risk-significant events may not be reported, especially if they were not anticipated. Moreover, it may be impossible to develop a precise definition of reportable precursors in relatively new or immature technologies and systems or in systems for which no quantitative risk analyses are available.

Conversely, setting the threshold for reporting too low runs the risk that the reporting system may be overwhelmed by false alarms, especially if the system design requires some corrective action or substantial analysis for all reported events. In addition, too low a reporting threshold can lead to a perception that the reporting system is of little value. These competing trade-offs can lead to errors, as shown in Table 1. Type I errors are reported events that do not pose a significant risk. Type II errors are events that do pose a significant risk but are not reported.

TABLE 1 Errors in Event Reporting

	Safety Significant	Not Significant
Event reported	True positive (the event is significant)	False positive (Type II error)
Event not reported	False negative (Type I error)	True negative (the event is not significant)

Finally, even reporting systems based on strict definitions of accident precursors with high thresholds for reporting may need a mechanism that allows for reporting new and previously unexpected precursors if they are judged to be severe. Sometimes, a single unrecognized or "hidden" flaw can render a technology much less safe than had been believed (Freudenburg, 1988), and precursor reporting systems are typically used for technologies in which unforeseen problems can have serious consequences.

THE OPPORTUNITY OF PRECURSOR MANAGEMENT

Programs for managing accident precursors have a number of benefits, as outlined by van der Schaaf et al. (1991). First, reviewing and analyzing observed precursors can reveal what can go wrong with a particular system or technology and how accidents can develop (modeling). For example, a precursor may reveal a previously unknown failure mode, which can then be incorporated into an updated model of accident risk. Second, because precursors generally occur much more often than accidents, analyses of accident precursors can help in trending the safety of a system (monitoring). For example, a precursor reporting system can provide evidence of improving or deteriorating safety trends and hence decreasing or increasing accident likelihoods. This information might not be apparent from sparse or nonexistent accident data. Trends in observed precursors can also be used to analyze the effectiveness of actions taken to reduce risk. Finally, and perhaps most important, precursor programs can improve organizational awareness (mindfulness) of safety problems (Weick and Sutcliffe, 2001). In organizations where actual accidents are rare, the dissemination of information on accident precursors can reduce complacency. Thus, the establishment of a precursor program may encourage an ongoing dialogue about safety in an organization, resulting in greater awareness of what can go wrong and greater willingness to discuss potential risks and safety hazards. Even if these discussions are not part of the formal precursor program, the more effective safety culture that they represent may still be a result of that program.

One way organizations seek to benefit from precursors is by analyzing near misses (sometimes referred to as near accidents, near hits, or close calls), fragments of an accident scenario that can be observed in isolation—without the occurrence of an accident. For a given accident scenario, near misses can and frequently do occur with greater frequency than the actual event (Bird and Germain, 1996). Several examples from the accident literature confirm this expectation, including the *Concorde* air crash (BEA, 2002), the London Paddington train crash (Cullen, 2000), and the Morton Salt chemical plant explosion (CSB, 2002); all three of these catastrophes were preceded by near misses, and some of the precursor events in the near misses were also parts of the eventual accident scenarios.

To organizations seeking to learn about potential accidents, near misses represent inexpensive learning opportunities for analyzing what can go wrong.

Near misses are especially important for organizations that have not experienced a major accident, because they enable these organizations to experience what March et al. (1991) refer to as "small histories"—or fragments of what might be experienced if an accident occurred. To benefit from near misses, organizations ranging from hospitals to manufacturing facilities and airlines to power plants, have set up management systems for reporting and analyzing near misses (see examples documented in this report and Barach and Small, 2000; Bier and Mosleh, 1990; Jones et al., 1999; van der Schaaf, 1992).

Analyses of accident precursor data can also be useful in conjunction with probabilistic risk analyses (PRAs). A PRA, also sometimes called a quantitative risk assessment or probabilistic safety assessment, is a method of estimating the risk of failure of a complex technical system by deconstructing the system into its component parts and identifying potential failure sequences. PRA has been used in a variety of applications, including transportation, electricity generation, chemical and petrochemical processing, aerospace, and military systems.

PRA methods make it possible to quantify the likelihood that each type of precursor will lead to accidents of different severities by assessing the conditional probability of accidents given certain precursors (Bier, 1993; Cooke and Goosens, 1990; Minarick and Kukielka, 1982). Such information can be helpful in prioritizing precursors for further investigation and/or corrective action. For an in-depth discussion of PRA, see for example Bedford and Cooke (2001) or Kumamoto and Henley (2000).

Precursor analyses have different strengths and weaknesses than PRAs and can, therefore, be used in conjunction with PRA models. PRA risk estimates are often heavily dependent on assumptions in the PRA model. For example, although every attempt is made to include important dependencies when they are recognized, a PRA may nonetheless incorrectly assume that two particular events are independent of each other. Because empirical data on observed precursors are relatively free of such assumptions, they can be used to assess the validity of those assumptions. Thus, if two events are positively correlated rather than independent, precursors involving both of them will tend to occur more often than predicted under the assumption of their independence, providing a potentially more accurate estimate of accident risks (and a check on the validity of the PRA model).

Other approaches have also been used to take advantage of precursor data. Automated surveillance systems, fault detection algorithms, and a variety of alarm systems are examples of systems that attempt to recognize precursors automatically. These methods have one common characteristic—they attempt to leverage precursor data to gain a better understanding of potential accidents.

Compared to purely statistical analyses of observed accident frequencies, near-miss analyses, PRA methods, and other precursor analyses can be viewed as examples of "decomposition" (i.e., breaking an accident scenario up into its component parts or building blocks). Forecasting expert J. Scott Armstrong of the Wharton School, University of Pennsylvania, notes that decomposition typically leads to better judgments, particularly in cases where uncertainty is high (as in the likelihood of an accident, where estimates can vary by orders of magnitude). Armstrong (1985) describes the following merits of decomposition:

> It allows the forecaster to use information in a more efficient manner. It helps to spread the risk; errors in one part of the problem may be offset by errors in another part. It allows the researchers to split the problem among different members of a research team. It makes it possible for expert advice to be obtained on each part. Finally, it permits the use of different methods on different parts of the problem.

Comparing Accident Analysis and Precursor Analysis

One of the most attractive aspects of precursor analysis is the abundance of precursor events compared to actual accidents (Bird and Germain, 1996). Thus, precursor data sets are often much richer than accident data sets. Analyzing precursor data can therefore reduce the uncertainty about the likelihood of an accident and lead to better decisions.

The committee believes that in many cases precursor events can be more effectively analyzed than accidents. After an accident, it may be difficult to determine what actually occurred for a variety of reasons: damage can be so severe that accident reconstruction may be inaccurate; the investigation may require too much time or money; legal and financial concerns may create disincentives that affect the investigation (e.g., individuals or organizations may be unwilling to disclose information that could increase their liability, or they may share information selectively); and witnesses may be unavailable. In contrast, when analyzing accident precursors, the system itself is usually intact, and stakeholders and witnesses may be more willing to report and share information about the event.

Comparing precursor analysis with accident analysis also reveals some of the challenges of benefiting from precursor information. Because precursors are likely to be numerous, resource limitations may make it impractical to investigate all of them to the level of detail that would normally be used in an accident investigation. Hence, thresholds are often set to select the precursors that are most indicative of system risk (Paté-Cornell, 1986). If a large number of precursors are considered important enough for analysis, they may be subjected to further prioritization and filtering.

Moreover, the potential for precursor events to develop into actual accidents might be unclear. As in any use of decomposition methods, the resulting model may not be entirely accurate (Bier et al., 1999); for example, there may be erroneous assumptions as to which additional events would be necessary to cause an accident given a particular precursor. In fact, non-accident precursors are inherently ambiguous (Bier and Mosleh, 1990) because they provide indications of system safety (e.g., the fact that an actual accident did not occur), as well as indications of risk (e.g., the fact that a precursor did occur). Thus, if a precursor occurs and no accident follows, some individuals may (correctly or incorrectly) conclude that the system is less prone to accidents than was initially believed, and there may be disagreements and debates about how seriously that precursor should be taken.

Because of their less dramatic end states, precursor events may seem less salient as lessons learned than accidents. For example, corrective actions developed in response to precursor data may be less persuasive and more open to question than corrective actions based on actual accidents (March et al., 1991). Because accidents are at least partly random, there is no guarantee that corrective actions adopted in response to even relatively severe precursors will actually prevent an accident. Decision makers may, therefore, pay less attention to precursors than to accidents, and it may be difficult to persuade them to make changes in technical or organizational designs based on observations of precursors.

Finally, legal concerns may compel an organization to analyze an accident thoroughly but may also inhibit the use of precursor data. For example, showing that an organization knew about a particular precursor but did not take corrective action could increase the organization's liability in the event of an actual accident. As a result, some organizations may be reluctant to establish formal precursor reporting programs; for example, they may rely on oral, rather than written, notification of observed precursors.

We can also compare the costs associated with precursor and accident analysis. Accidents can have a number of direct costs, such as medical expenses, costs associated with employee convalescence, and equipment damage. In contrast, precursor events may have minimal if any direct costs. Accidents also have a number of indirect costs that may far outweigh the direct costs. Typical indirect costs include lost production, a drop in employee morale, scheduling delays, additional hiring/training, legal costs, and the costs of implementing corrective actions. After a precursor event, many of these indirect costs may not apply (e.g., there may be no lost production) or may be lower than if an actual accident had occurred.

From this comparison, one might wonder if implementing a precursor analysis program can be more cost effective than assuming the risks and costs of the accident the program is intended to prevent. To the committee's knowledge, no comprehensive cost-benefit analysis of precursor analysis programs has been conducted. Nonetheless, the committee firmly believes that precursor programs can be, and often are, cost effective. That is, the costs associated with achieving risk reduction through a precursor program are far lower than the risk-adjusted costs assumed when no such program is in place and precursors are not systematically analyzed.

Encouraging the Use of Precursor Analysis

The relatively high frequency and low cost associated with precursor events suggest that many industries could benefit from using precursor analyses to reduce the risk of accidents. Perhaps not surprisingly, industries that have traditionally sought to benefit from precursor analysis (e.g., aviation, aerospace, nuclear power, and the chemical process industry) are subject to accidents that can be so severe, but also so infrequent, that the advantages of precursor analysis are especially attractive.

One factor that seems to be essential for the adoption of precursor programs is the active engagement of companies—a company must "own" a precursor program. Thus, an organization must have leadership and a "safety culture" that can support such a program. The concept of a safety culture was developed by the International Atomic Energy Agency in the analysis of contributing factors to the Chernobyl disaster (Wiegmann et al., 2002). Although there are a number of industry-specific definitions of safety culture (see Wiegmann et al., 2002, for several examples), Pidgeon (1991) provides one encompassing definition:

> [A safety culture is] the set of beliefs, norms, attitudes, roles, and social and technical practices that are concerned with minimizing the exposure of employees, managers, customers and members of the public to conditions considered dangerous or injurious.

Carroll and Hatakenaka (2001) describe an example of a plant, the Millstone Nuclear Power Station, in New London, Connecticut, that underwent an organizational shift and became a safety-conscious work environment that exhibited many of the characteristics associated with a healthy safety culture. In 1996, the Millstone Nuclear Power Station was featured in a *Time* magazine cover story as a rogue utility that cut corners and intimidated or fired employees who raised safety concerns (Pooley, 1996). The U.S. NRC placed Millstone on a watch list of plants receiving additional regulatory attention, and, following a shutdown of the plant's three units, ordered that all three demonstrate that they were safe and in compliance with license and regulatory requirements prior to restarting.

In an effort to address shortcomings in compliance and safety, Northeast Utilities (Millstone's owner) changed the top leadership of its nuclear program and brought in Bruce Kenyon to be CEO of Northeast Nuclear Energy Company. Carroll and Hatakenaka (2001) describe how Kenyon engineered an organizational transformation. Afterward, instead of suppressing the sharing of safety-related concerns, leadership of the company considered it essential that safety concerns be shared among employees and management. Some of the key changes included: the dismissal or demotion of senior managers who were identified by their peers as underperformers; the hiring of new managers to run the employee safety-concern program; the creation of formal structures and forums for two-way communication for the sharing of safety-related information between employees and management; and the hiring of third-party consultants to oversee and monitor the effectiveness of instituted changes.

Carroll and Hatakenaka's (2001) account of Millstone's transformation underscores that leadership is essential but not the sole component of an effective safety culture; all members and strata of the organization must embrace the safety culture. Nonetheless, if the parent company's leadership had not embraced the sharing of safety-related concerns and instituted changes to enable this sharing, it appears unlikely that Millstone would have been able to transform itself.

Leadership may be even more important in organizations and industries with less regulatory oversight or where safety reporting is voluntary. In such organizations, a culture and leadership that encourage reporting may be one of the few compelling reasons for employees, contractors, and front-line managers to share safety concerns and, potentially, information regarding precursors to accidents.

The private sector, industry associations, government, and third parties can all play a role in helping organizations understand and manage their risk exposures through the sharing of risk-related information and precursor analysis. Economic and regulatory mechanisms can provide incentives for organizations or companies to institute precursor analysis programs.

Some regulatory agencies use command-and-control regulation to mandate the reporting of certain types of precursors (e.g., the Licensee Event Reports mandated by the U.S. NRC in Code of Federal Regulations 10CFR50.83). Other organizations have voluntary programs, such as the Aviation Safety Action Program (discussed below), that protect individuals who report precursors from sanctions provided that certain "cardinal rules" are followed. Adhering to nonpunitive guidelines (under which individuals are not punished for reporting events in which they were involved) helps to build and maintain trust, although there is generally a threshold above which some type of punishment may apply. For example, incidents that involve clear violations, such as criminal or malicious behavior, are typically managed separately from precursor programs to avoid protecting individuals who have committed such violations.

Other incentives to encourage precursor management could include monetary or other rewards for companies that institute programs to identify and collect data on precursor events. For example, insurance premiums could be reduced for organizations that try to reduce their risk exposure through the systematic use of precursor information (Kunreuther et al., 2003). In lieu of involvement by regulatory agencies, third parties, such as trade organizations, insurers, accrediting bodies, and comparable companies, could inspect companies to determine whether they have effective and appropriate precursor programs in place (Er et al., 1998; Kunreuther et al., 2002).

Legal safeguards could also be used to protect individuals and companies that collect and share information about risk. Under current law, precursor reports generated prior to an accident are often considered discoverable evidence after an accident. This may deter companies from soliciting and collecting reports about safety problems, and some industries have taken steps to insulate reporters of safety problems from liability.

LEARNING FROM PAST EXPERIENCE

The loss of the space shuttle *Columbia* and other major events (such as the terrorist attacks of September 2001) and recent lapses in safety (such as the serious corrosion problems discovered at the Davis-Besse nuclear power plant in Ohio in 2002 and the major blackout in the eastern United States in August 2003) have raised questions about how organizational structures and cultures can learn from precursors. These events have raised issues about how knowledge can be disseminated and applied throughout an organization; the feasibility and challenges of transferring precursor approaches from one industry to another; and the potential transferability of precursor approaches to problems outside the area of technological accidents.

The Space Shuttle *Columbia*

The *Columbia* accident occurred about seven months prior to the workshop. The signals leading up to the accident and how NASA managed them were analyzed extensively by the *Columbia* Accident Investigation Board (CAIB, 2003). Like the analyses of many other accidents, the CAIB study of the *Columbia* accident (mission STS-107) revealed a number of warning signals suggesting that the likelihood of an accident was greater than NASA had perceived at the time.

Considerable analysis by the CAIB (2003) addressed what sociologist Diane Vaughan calls "the normalization of deviance" (Vaughn, 1997). The CAIB report concluded that, although certain precursor events in missions prior to STS-107 had indicated problems, their continued occurrence without resulting in accidents had led to a misperception they were consistent with normal operation. In other words, precursors were initially considered warning signals, but over time were no longer considered indicative of serious risks. The CAIB report argued that NASA had thus "normalized" precursor events that today are generally believed to have been the direct cause of the orbiter loss.

The direct cause of the accident appears to have been insulating foam detached from the external tank striking the left wing of *Columbia* during the orbiter's ascent and piercing the orbiter's thermal protection system. During reentry into the Earth's atmosphere, hot plasma gases then entered the orbiter and disintegrated the orbiter's internal structure (CAIB, 2003). Debris strikes that had not penetrated the thermal protection system had been well documented in previous missions and had been carefully monitored and analyzed. Debris strikes resulting from detached foam had been observed in 65 of the 79 missions for which photographic imagery was available.

In fact, Paté-Cornell and Fischbeck (1993) had undertaken PRA studies to analyze the case of foam becoming detached from the external tank, hitting the tiles of the orbiter, and causing enough damage to the thermal protection system to result in "burn-through" during reentry. They concluded that the likelihood of this event was sufficiently high to merit some attention to this problem.

The debris strike on STS-27R (on December 2, 1988) was similar to the eventual failure of the *Columbia* on mission STS-107, but the CAIB noted that during STS-27R, the orbiter had been inspected and managed much more diligently than during STS-107. The CAIB concluded that NASA's perception of the severity of debris strikes had changed between missions STS-27R and STS-107: "NASA engineers and managers increasingly regarded the foam-shedding as inevitable, and as either unlikely to jeopardize safety or simply an acceptable risk." The CAIB report concluded that the shuttle program lacked the "institutional memory" to benefit from the lessons of STS-27R (CAIB, 2003). This finding demonstrates how changes in organizational culture can affect the way precursors are perceived and managed.

INTRAORGANIZATIONAL SHARING AND ANALYSIS OF PRECURSOR INFORMATION

Some researchers believe that certain complex, tightly coupled, high-hazard organizations routinely maintain better than expected levels of safety and reliability. These are generally referred to as "high-reliability organizations" (HROs). Examples of HROs include nuclear power plants (Bourrier, 1996; Marcus, 1995), air traffic control systems (LaPorte, 1988; LaPorte and Consolini, 1998), and aircraft carriers (LaPorte and Consolini, 1998; Roberts, 1990; Rochlin et al., 1987; Weick and Roberts, 1993)

Researchers on the cultures, structures, and processes of HROs have postulated that one of the defining characteristics of HROs is a high sensitivity to things that can go wrong. HROs are believed to have organizational cultures that encourage "a rich awareness of discriminatory detail and facilitates the discovery and correction of errors capable of escalation into catastrophe" (Weick et al., 1999).

One factor that contributes to greater sensitivity and attentiveness to precursors in HROs is transparency, that is, an environment conducive to the free flow of information about potential risks. In some organizations, such as air traffic operations, in which constant communication reinforces confidence in the integrity and status of operations, information is shared almost continually (Rochlin, 1999). Data may also be exchanged on a more occasional basis through informal channels that encourage discussions of risks and lessons learned at all levels of an organization (Roberts, 1990). Either way, the important point is that the climate created makes it easy for information about problems to be brought to the attention of key decision makers, including front-line personnel and senior managers.

It is important to keep in mind that attentiveness to precursors is not the only characteristic of an HRO. Organizations may exhibit a high sensitivity to precursors but fail to achieve high reliability because they do not have key characteristics of effective safety management. As Westrum and Adamski (1999) and Dowell and Hendershot (1997) point out, the search for errors can sometimes increase system risk if intended safety improvements inadvertently create more risk-prone systems. As Rochlin (1999) observes, "the search for safety is not just a hunt for errors."

INTERORGANIZATIONAL SHARING OF INFORMATION

The management and exchange of information pertaining to risks beyond a single organization is an important issue associated with precursor management. Organizations can be deluged with information from internal and external sources, which can make filtering and recognizing problem areas and recognizing precursors to accidents more difficult. This, in turn, makes it more difficult to determine which information should be shared outside the organization. Even for precursors that are recognized, concerns about releasing proprietary knowledge, tarnishing a firm's image, or incurring legal recriminations may discourage information sharing.

Sharing of information across organizations is important because many hallmark accidents that have drawn attention to the importance of precursor management were preceded by similar but non-catastrophic precursor events in other organizations. Because of a lack of effective information exchange, the organization that experienced the eventual accident was often unaware that others had learned from and acted on related precursor events.

This was the case in the Three Mile Island (TMI) accident, in which one of the factors in the partial core meltdown was a pressure relief valve that was stuck open, leading to confusion and misinterpretation in the plant control room. A similar event in which signals from a stuck relief valve had been temporarily misinterpreted had occurred at the Davis-Besse nuclear power plant in Ohio a year-and-a-half earlier. Fortunately, the progression of the Davis-Besse scenario had been halted, and an accident at that plant was averted. Although the Davis-Besse management had documented the event and learned from it internally, the information had not been shared with anyone outside the plant. Thus, management at TMI was not able to benefit from the experience (Chiles, 2002).

A similar situation led to the development of the ASRS (Aviation Safety Reporting System) in the aviation industry. On December 1, 1974, TWA Flight 514 was inbound to Dulles Airport near Washington, D.C. During the descent, the flight crew misunderstood the approach instructions and descended prematurely to the final approach altitude. The premature descent, coupled with limited visibility due to inclement weather, significantly contributed to the pilots flying the aircraft into a mountaintop, killing everyone on board. During the National Transportation Safety Board's accident investigation, a disturbing finding emerged. Six weeks prior to the accident, under similar conditions, a United Airlines flight crew had experienced a similar misunderstanding and had narrowly averted hitting the same mountain. After landing, the crew had reported the near miss to their company's new internal reporting program, and an alert had been issued to all United Airlines pilots about the potential hazard. Because there was no established mechanism for sharing this information externally, the crew of TWA 514 was unaware of the hazard (ASRS, 2001).

Research suggests that transparency and the free flow of information should ideally extend to observers *outside* of an organization (i.e., "institutional permeability"). The need is illustrated most vividly when the absence of institutional permeability contributes to disasters. For example, Turner and Pidgeon (1997) discuss cases in which "individuals outside the principal organizations . . . had foreseen the danger which led to the disaster, and had complained, only to meet with a high-handed or dismissive response." The examples include a mine-tailings landslide that killed 144 people and a rail-crossing accident that killed 11 people. Related issues are addressed by Lodwick (1993) and Martin (1999).

Chess et al. (1992) note that "organizations can develop systems to amplify the concerns of those outside the plant so that these voices can be heard easily by personnel inside the plant who have the capability to reduce risk"; they also describe how this was achieved by a small chemical manufacturer through the implementation of "an exemplary risk communication program." Although concerns about protecting proprietary information are valid, some level of institutional permeability (especially receptiveness to concerns raised by "outsiders") can expand the range of information available to an organization and can counteract complacency and the normalization of deviance.

SAMPLE INDUSTRY APPROACHES

A number of industries have implemented programs for taking advantage of precursor information, several within the past few years. They include the Accident Sequence Precursor (ASP) Program and the Institute for Nuclear Power Operation's Significant Event Evaluation and Information Network in the nuclear industry, the ASRS in the aviation industry (DOT-FAA, 2002), site-specific and company-specific near-miss programs in the chemical industry (van der Schaaf, 1992), the U.K. rail industry's confidential reporting systems (CIRAS, 2003), voluntary reporting programs for maritime safety (BTS, 2002a), surveillance systems to detect adverse drug events in health care (Kilbridge and Classen, 2002), national voluntary reporting systems in health care (IOM, 2000), and motor vehicle safety programs defined under the TREAD Act (DOT-NHTSA, 2002).

To illustrate the differences among these approaches, several methods of collecting and analyzing precursor data are highlighted below. These descriptions are not intended to be representative of all approaches used in a given industry, and the committee does not endorse one approach over another.

Accident Sequence Precursor Program

The ASP Program, overseen by the U.S. NRC, analyzes and disseminates findings from potential precursor events at U.S. commercial nuclear power plants. This nationwide precursor program overseen by a federal agency is discussed in more detail in the paper by Martin Sattison (p. 89 in this volume). The ASP Program was initiated several decades ago, following publication of the first PRAs of nuclear power plants to analyze precursors to a potentially catastrophic core meltdown by (USNRC, 1978):

- quantifying and ranking the safety significance of events at operating reactors
- determining the generic implications of these events
- characterizing risk based on those events
- providing feedback for operators of other plants to learn from these experiences

The ASP Program defines an accident sequence precursor as an operational event or plant condition that is an element of a postulated accident sequence that could lead to inadequate core cooling and hence to core damage. The precursors analyzed in the ASP Program are selected primarily from Licensee Event Reports that must be submitted to the U.S. NRC by plant licensees. Each event is reviewed to determine its severity and relevance to safety. Accident precursors estimated to have a conditional core damage probability greater than 1.0×10^{-6} (greater than a one in a million chance of resulting in core damage) are selected for further analysis (Johnson and Rasmuson, 1996; Reisch, 1994).

Aviation Safety Action Programs

Aviation safety action programs (ASAPs) are airline-initiated programs that encourage employees to voluntarily report safety information that may be critical to identifying potential accidents (DOT-FAA, 2002). ASAPs are based on memorandums of understanding (MOUs) between airlines (or repair stations), the FAA, and applicable third parties representing employees, such as labor associations. Although ASAPs are carrier operated, the programs must adhere to federal guidelines, and information is shared between the carriers and the FAA. In a recent advisory circular, the FAA states (DOT-FAA, 2002):

> The objective of the ASAP is to encourage air carrier and repair station employees to voluntarily report safety information that may be critical to identifying potential precursors to accidents. The Federal Aviation Administration has determined that identifying these precursors is essential to further reducing the already low accident rate.

Although ASAPs are company-administered programs, all signatories to an MOU must adhere to its provisions in the execution of the program. ASAP guidelines have been updated periodically after analyses of demonstration programs and as more companies have developed their own ASAPs (DOT-FAA, 1997, 2000, 2003). Although reports are managed internally, the information is shared with the FAA and throughout the industry when warranted.

Each ASAP has an event review committee (ERC) that evaluates whether submitted reports should be included in the ASAP program. Members (and alternates) of an ERC are designated representatives of the FAA, the certificate holder (i.e., an airline or repair station), and a representative of a third party, such as an employee union. ERCs have five specific responsibilities (DOT-FAA, 2002):

1. Reviewing and analyzing reports submitted to the program.
2. Determining whether reports qualify for inclusion in the program.
3. Identifying actual and potential safety issues from the information in the reports.
4. Proposing corrective actions to remedy identified safety concerns.
5. Following up on ERC recommendations for corrective actions to assess whether they have been satisfactorily accomplished.

Several demonstration programs initiated after a 1997 advisory circular (DOT-FAA, 1997) have engaged employees in discussing safety issues. Among these programs are the USAir Altitude Awareness Program, the American Airlines Safety Action Partnership, and the Alaska Airlines Altitude Awareness Program. Since their inception, more than two dozen ASAP programs have been established (DOT-FAA, 2002). To encourage wider participation by carriers, President Clinton announced that ASAPs would be part of a national effort to reduce aviation accidents (White House, 2000).

ASAPs have been promoted because they encourage aviation employees to report safety problems quickly (DOT-FAA, 2002). The programs stress implementation of corrective actions over punishment and discipline, although the FAA can prosecute cases involving egregious acts (e.g., substance or alcohol abuse or the intentional falsification of information). ASAPs provide previously unavailable information rapidly and directly from those responsible for day-to-day aviation operations. These programs are expected to lead to improvements in FAA management of the National Aerospace System, airline flight operations and maintenance procedures, pilot-controller communications, human-machine interactions and interfaces, and training programs, ultimately helping to meet the

FAA's goal of reducing the accident rate for commercial aviation by 2007 (White House, 2000).

Adverse Drug Events

Programs to detect potential and actual adverse drug events (ADEs) in health care are examples of how precursors can be actively and automatically monitored and how work processes can be structured around precursor detection. ADEs, events in which patients are harmed as a result of drug interventions, are some of the most frequent negative outcomes in health care, and their cumulative effects are enormous. Every year, an estimated one million serious medication errors are made in hospitals (Birkmeyer et al., 2000). Two well known cases of fatal ADEs are the deaths of Betsy Lehman (a health care reporter for the *Boston Globe*, who died of a chemotherapy overdose after being given four times the normal dosage over a four-day interval [Cook et al., 1998]) and Libby Zion (an 18-year-old woman who died when she took a prescribed drug that had a known, potentially fatal interaction with an antidepressant she was also taking [Asch and Parker, 1988]). Health care institutions have recently shown a good deal of interest in creating surveillance systems to monitor ADEs (see, for example, Bates et al., 1999, and Kilbridge and Classen, 2002).

ADEs can occur for a wide variety of reasons (Classen, 2003). Allergic reactions, drug-drug interactions, incorrect dosage prescriptions, incorrect dosage administration, and unintended repeated dosages are a few common ADEs. Although voluntary reporting systems encourage the reporting of these events or their precursors, many ADEs appear to go unreported (O'Neill et al., 1993). An alternative approach is to implement surveillance systems that automatically monitor for precursor events and to establish work processes to ensure that when an incident is detected, the impending accident is averted.

An example of the latter approach is prescription error-detection software, which is often integrated into computerized physician order entry (CPOE) systems used for ordering medications. Once a surveillance system has been implemented, a wide variety of precursors to ADEs can be detected, and potential harm to patients can be averted. For instance, if a doctor mistakenly orders penicillin for a patient who is allergic to it, an alert automatically informs the doctor of the precursor event.

When implemented successfully, surveillance systems have been shown to decrease ADEs dramatically (Bates et al., 1998, 1999; Evans et al., 1998). Based on the potential of surveillance systems to improve safety, the Leapfrog Group (a coalition of large health care purchasers that seeks to align health care purchasing with health care safety) has encouraged hospitals to implement CPOE systems. To meet Leapfrog's CPOE standard, hospitals must satisfy the following requirements (Leapfrog Group, 2003):

- Ensure that physicians enter hospital medication orders via a computer system that includes error-prevention software.
- Demonstrate that the inpatient CPOE system can alert physicians to at least 50 percent of common, serious prescribing errors using a testing protocol now under development.
- Require that a physician electronically document the reason for overriding an interception prior to doing so.

An automated surveillance approach could also potentially be applicable to other industries. In fact, a wide variety of alarm systems can be considered surveillance systems for detecting precursors to accidents. For example, near midair collisions and trains passing a red signal (indicating danger), both of which are generally considered precursors to accidents, can be automatically detected.

Surveillance systems have certain advantages over voluntary reporting systems. First, surveillance systems can often be built into work-flow processes so that precursors that might otherwise progress to accidents can be halted through detection and alerts. In addition, these systems frequently yield higher reporting rates than voluntary reporting systems and sometimes even encourage individuals to submit more voluntary near-miss or safety-related reports. However, surveillance systems also have some drawbacks. For instance, they may not capture all types of precursors because they generally detect only unambiguous signals that are known to have the potential to progress to accidents and that can be readily monitored. In addition, surveillance systems can create new, unexpected problems. For instance, if alerts are triggered too often, people may disregard them.

FINDINGS AND RECOMMENDATIONS

These findings and recommendations are based on surveys of the literature by National Academy of Engineering staff and the project committee, committee meetings, workshop presentations, feedback from workshop participants, and the workshop papers reproduced in this report. The recommendations are intended to help organizations design, refine, and oversee precursor programs and to help government agencies encourage the use of precursor data in a range of domains. In keeping with the cross-industry focus of the study, the recommendations are not industry specific. The findings and recommendations are presented in five sections—opportunity, precursor management, organizational commitment, engaging industry, and engaging government.

Opportunity

Finding 1. The collection, filtering, and analysis of accident precursor data, followed by the implementation of corrective actions, can improve reliability and safety.

There is ample evidence showing that improvements have resulted from precursor-type programs. In aviation, for instance, a variety of precursor programs have led to improvements in safety. Flight operational quality assurance (FOQA) programs, in which flight data are routinely analyzed regardless of whether an incident was observed or reported, have identified a number of potential precursors and led to the adoption of new safety measures. These include modifications of pilot training, revisions to or renewed emphasis on standard operating procedures, equipment fixes, and the issuance of alerts to pilots regarding potential hazards (GAO, 1998). The Flight Safety Foundation's publication, *Flight Safety Digest*, shows that other aviation safety reporting and sharing platforms, including ASAPs, ASRS, and the Global Aviation Information Network, also frequently identify precursors and support analyses of precursor events (DOT-FAA, 2002). Studies of other industries also cite safety improvements after the institution of precursor programs (see examples in the papers by James Bagian [p. 37] and Dennis Hendershot [p. 103] in this volume).

This finding does not explicitly address the cost effectiveness of precursor programs. However, as indicated earlier, continued major lapses in safety management (such as the loss of the space shuttle *Columbia*, the corrosion problems discovered at the Davis-Besse nuclear power plant in 2002, and the August 2003 blackout) suggest that we are far from the point of diminishing returns on investments in safety.

Recommendation 1. Organizations involved in operations with significant safety and reliability concerns should evaluate the opportunities for risk reduction through precursor analysis programs.

Precursor Management

The effective management of precursors, near misses, and close calls poses a number of challenges. Managing a single incident involves recognizing that a precursor has occurred, ensuring that the event is reported, and analyzing the event to assess its causes and identify possible corrective actions. Managing an entire precursor program requires identifying the types of precursors to be reported, prioritizing and filtering observed incidents (e.g., deciding which precursors justify reporting, which reports justify further analysis, and which analyses justify corrective actions), and deciding which reports to disseminate and which corrective actions to implement on an organizational scale.

The following findings address specific issues associated with the management of accident precursors. They are not intended to be comprehensive, and some aspects of precursor management (such as root-cause analysis, discussed by William Corcoran, p. 79 in this volume) are not addressed here.

Finding 2. Effective precursor management programs include clear definitions of risk, risk-reduction objectives, and the types of precursor data needed for risk management.

Defining Precursors

The range of precursors reported depends on how precursors are defined. Definitions vary from highly specific criteria (such as exceeding a specific quantitative threshold) to broad definitions that encompass a wide range of events and circumstances. Definitions of near misses and close calls can also vary from one industry or setting to another.

Designers and managers of precursor programs may assume that participants know what types of events to report and that they will recognize them when they occur. However, even highly knowledgeable individuals can have different views of the meaning of accident precursors, which can substantially affect the range of incidents reported. Phimister et al. (2003) cite examples from the chemical industry of personnel identifying precursor events that would have been of interest to management but not reporting them because they did not match the stated definition of the precursor program.

Recommendation 2. Precursor programs should define the precursors of interest in a way that is readily understandable to everyone expected to report a precursor, close call, near miss, or other safety-related occurrence.

Finding 3. The expected operation of a technology is not always characterized in a way that makes deviations readily apparent. This can result in precursors going unreported.

Although it is not always possible to distinguish between normal and abnormal operations, distinguishing precursor events based on a defined, ideal mode of operation has several advantages. First, if participants in precursor programs have a clear understanding of the standards of operation, they can compare an observed incident with the standards to determine if the deviation is significant. Second, a clear understanding of ideal operation can provide a basis for deciding whether a corrective action is necessary and, if so, which action to take. Third, explicit contrasts between precursors and the standards of operation can help in the prioritization of observed precursors.

Defining ideal operations involves not only knowing about the operation of the system in question, but also making value judgments about the range of acceptable deviations. This requires the identification of a consistent threshold between ideal and abnormal operations. Although some deviations from ideal operation may be considered acceptable (and may, in fact, be unavoidable in some situations), Vaughan (1997) has illustrated the risks associated with the normalization of deviance. Therefore, there should be a high "safety margin" in evaluating the risks posed by deviations. Deviations that are judged to be unacceptable after careful scrutiny should trigger corresponding contingency responses.

Recommendation 3. Activities with potentially significant risks should be subjected to an appropriate level of hazard analysis, which should then be used to help identify and define precursor events of concern.

Reporting Precursors

Finding 4. Barriers to reporting precursor events include a variety of factors: fear of blame for an event; reluctance to report a coworker's failure; concerns about liability; and lack of time to complete reports.

Precursor events that do not result in damage or loss, are witnessed by only a few people, or that cannot be readily monitored by a surveillance system can be difficult to capture in a reporting system. For management to learn of such events, the workforce must be actively engaged in the program. Christopher Hart outlines a number of legal and political barriers that can impede the reporting of potential errors to management or regulatory authorities, including (p. 147 in this volume):

1. The belief that an individual may be held responsible for a precursor event that he or she reports.
2. The potential for criminal prosecution of the individuals involved in an event.
3. The possibility that the information could be disseminated to the public.
4. The possibility that the information could be used in civil litigation proceedings.

Others have cited additional barriers to reporting, including lack of confidence that a report will result in safety improvements and lack of time to complete the report and still complete other tasks (Bridges, 2000). Management must develop strategies to overcome such barriers.

Recommendation 4. Organizations that implement precursor management systems should ensure that the work environment encourages honest reporting of problems as part of a positive safety-improvement culture.

Prioritizing Precursors and Disseminating Precursor Information

Finding 5. Organizations considering or implementing precursor programs face a variety of challenges, including filtering and prioritizing reports for effective analysis and identifying sound risk-reduction responses to observed precursors.

Programs that motivate individuals to report precursors face other challenges, such as how to manage the reported information effectively. If only a few reports are submitted, they can all be analyzed and disseminated to the relevant parties (as is typically done for serious accidents). However, if a large number of precursor reports are submitted, resource constraints may make it difficult to analyze all of them, and it may be impractical to share information about all reported events with everyone participating in the program. For example, ASRS receives about 2,900 reports a month, only 15 to 20 percent of which are logged because of resource constraints (Strauss and Morgan, 2002).

Prioritizing precursor events once they have been reported can also be a challenge. A number of approaches are currently used to prioritize precursors. In some programs, one or more individuals involved in the program simply screen precursor events and prioritize them subjectively. Sometimes, a database of historical events and precursors is used for trending purposes (e.g., to identify increasing or decreasing rates of particular types of precursors over time). In addition, mathematical modeling can be used to assess the probability of an accident conditional on a given type of precursor—as a measure of precursor severity, for example. PRA can be used to estimate the likelihood of accidents based on precursor information and to reduce uncertainties about accident risk. Delphi approaches can also be used to solicit and aggregate expert information on the likelihood of accidents.

Recommendation 5. Organizations should link precursor programs to the hazard assessment methodology used to manage safety and reliability, thereby developing a basis for setting priorities and using precursor information to establish measurements for improvements in risk.

Organizational Commitment

The ability to leverage precursor information to reduce risk exposure depends heavily on organizational endorsement, commitment, and leadership. Organization leaders must be involved in the development and implementation of precursor programs and must have a clear understanding of each program's structure, merits, and potential vulnerabilities.

Finding 6. Each organization has its own management structures, history, and culture, which are integral to both its safety philosophy and the role of precursor programs as part of the organization's commitment to safe, reliable operation.

The design of a precursor program must be sensitive to the characteristics of the particular situation, such as management structures, industry and organizational history, government and labor relations, the regulatory environment, legal considerations and constraints, the financial health of the industry and organization, and public perceptions of the risks posed by the industry in question.

To ensure continued participation, precursor programs must also lead to demonstrable improvements in safety. Because improvements resulting from precursor programs may not be readily visible to the casual observer, they should be audited and evaluated in terms of both risk reduction and cost effectiveness, and the resulting information should be shared with the people expected to participate in the program to encourage them to continue their participation. Evaluating whether safety improvements achieve the desired objectives requires organizational and management commitment to the program, as well as adequate resources.

Recommendation 6. Precursor programs should be implemented with the commitment of management at all levels, and measurable safety improvements attributable to the program should be publicized.

Engaging Industry

Finding 7. Many precursor events (and major accidents) occur in the private sector. Therefore, to reduce accident rates through precursor management, the private sector must be actively engaged in identifying and managing precursor events.

Although an increasing number of companies in high-hazard industries (i.e., industries that may experience catastrophic events) have initiated precursor or near-miss reporting programs, the committee believes this represents only a small fraction of the companies that could benefit from such programs. The committee encourages companies that do not have programs in place to examine industry best practices and implement programs suited to their needs and the hazards they face.

Recommendation 7. Companies in high-hazard industries should institute and/or maintain formal precursor programs for the collection, analysis, and sharing of risk-related information.

Finding 8. In some cases, channels for communicating risk-related information among companies in high-hazard industries are weak or nonexistent.

Many companies have valid concerns about sharing information, such as concerns about releasing proprietary information and/or the legal implications of sharing information. As a result, important information may either not be shared or may be shared only after it has been stripped of essential facts, so that it is of relatively little use to the recipient.

Participation by multiple parties in information sharing often amplifies the benefits derived from the information, especially when the parties face common risks. Hence, the committee encourages companies to work to overcome the barriers and develop novel approaches to sharing risk-related information. For instance, in a regulated industry, a private third party could play the role of honest broker, instead of a government agency, with government approval of the overall approach. A similar model is already being used in the chemical industry, where a number of chemical companies participate in the Process Safety Incident Database maintained by the Center for Chemical Process Safety (CCPS). The CCPS (a division of the American Institute of Chemical Engineers) collects, de-identifies, and shares anonymous information about accidents, incidents, and near misses with participating companies (Kelly and Clancy, 2001).

Recommendation 8. Companies in high-hazard industries should develop strategies for sharing risk-related information with other companies, when possible, as well as with other plants and facilities within their own companies, and should work to make proprietary information "shareable" between companies.

Finding 9. Greater cross-industry sharing of risk-related research, experiences, and practices could be widely beneficial, as evidenced by the cross-industry learning experienced at the workshop.

The advance of precursor practices and research requires open channels of communication—not only among the facilities of a single company or among firms in the same industry, but also among industries. It was evident at the workshop that industries have much to learn from each other and that obstacles in one industry might be overcome by leveraging the research and practices of other industries. More cross-industry sharing would encourage both research and the conversion of research results to reliable, effective practices. Cross-industry sharing could be facilitated by bringing together members of high-hazard industries regularly to discuss risk-related issues. This could be done by trade organizations, the National Academies, the Society for Risk Analysis, the Public Entity Risk Institute, and/or government bodies.

Recommendation 9. Organizations should support and participate in cross-industry collaborations on precursor management and research.

Engaging Government

Even though government institutions are already engaged in facilitating the reporting and analysis of precursors, the committee believes that government could do more to foster the cross-company and cross-industry sharing of information. However, government actions must be carefully considered to ensure that they encourage rather than discourage individuals and organizations from participating in precursor identification and management programs.

Finding 10. Existing regulatory models for using precursor data are potentially applicable to multiple industries.

Government agencies seeking to leverage precursor information in an industry should consider adapting approaches that have already been developed for other industries. For example, analogous versions of the ASAP and ASRS models have been developed for industries other than aviation. In the ASAP model, each company collects and manages near-miss and precursor data in parallel with other companies using similar data-collection methods. Phimister et al. (2003) and Barach and Small (2000) discuss similar reporting systems in the chemical and health care industries, respectively. In the ASRS model, a third party (in this case, NASA) is endorsed by the regulatory agency as an honest broker. The Department of Veterans Affairs uses a similar reporting system in health care settings.

Transferring precursor program models from one industry to another must be done carefully, however. Workforces may have different cultures that affect the acceptability of particular models; stakeholders may have different relationships; issues of proprietary information may impede the transfer of safety-sensitive information; and legal issues may hinder the sharing of information. Finally, incentives for sharing information about risks may differ from one industry to another. Steps that can be taken to encourage the adoption of precursor programs include providing economic incentives for information sharing, aligning market mechanisms to encourage precursor management (e.g., through reductions in insurance premiums), and third-party inspections of corporate risk-management programs (Carroll and Hatakenaka, 2001; Kunreuther et al., 2002).

Recommendation 10. Government agencies overseeing high-hazard industries or technologies that do not have a cohesive strategy for managing precursor information should develop an initial agency policy on precursor management to initiate a dialogue on how precursors can and should be managed.

The committee notes that some industries and agencies have already initiated activities consistent with this recommendation. For example, a white paper prepared by the Volpe Center (2003) served as the basis for a discussion at a railroad industry workshop held in 2003. The paper and workshop helped initiate an industry dialogue to evaluate how precursor information is currently used in the industry and how it could be used more effectively to improve railroad safety. In addition, as part of the Safety Data Initiative at the Bureau of Transportation Statistics, working groups have been charged with collecting better data on accident precursors and expanding the collection of near-miss data to all modes of transportation (BTS, 2002b).

Finding 11. There is already an ongoing research agenda in precursor analysis and management.

The committee believes that further research on precursor management would lead to higher levels of system safety. Given the number and severity of technological accidents in the past two decades, research should be considered a high priority for agencies that regulate high-hazard industries. The source(s) and amount of funding for such research will vary from one industry to another.

Because many disciplines in engineering, physical sciences, and social sciences can contribute to precursor analysis and management, and because the research needs vary from one industry to another, it is difficult to prioritize research topics. However, areas of general interest that may benefit precursor management programs might include: the identification of trends in large amounts of statistical data; the design of fault-tolerant systems; human factors analysis; the design of human-machine interfaces; team dynamics in safety-critical system operations; and organizational learning and leadership.

Research topics directly usable in precursor programs might include: data acquisition methods; improved fault-detection algorithms; risk modeling and trending methods; the relative effectiveness of alternative regulatory frameworks for precursor reporting and management; industry epidemiological analyses; and strategies for engaging large organizations in risk management. Academia, industry, government, and collaborative public-private projects could all be involved in research on these topics and other challenges identified in the papers in this report.

The committee also believes that basic research on precursor management would benefit numerous industries. Some of the most effective practices in precursor management are summarized in this report, but there are still significant uncertainties about the effectiveness of different approaches—partly because of insufficient scientific evaluations of precursor management methods. For example, basic scientific research could compare the merits of voluntary and mandatory reporting systems or quantify the decrease in system risks affected by precursor programs (e.g., using PRA or industry epidemiological analysis). The committee encourages the National Science Foundation and the mission agencies to support basic research in these and related areas.

Recommendation 11. Mission agencies with discretionary research budgets should support precursor-related research and pilot studies relevant to their respective missions. In addition, funding agencies and foundations should support basic research on using accident precursors in risk management programs and the characteristics of effective precursor information management.

CONCLUSION

The practice of searching for and learning from accident precursors is a valuable complement to other safety management practices, such as sound system engineering, adherence to standards, and the design of robust, fault-tolerant systems. Maintaining safety is an ongoing, dynamic process that does not stop when a technology has been designed, built, or deployed. Despite the best engineering practices, and despite strict adherence to standards and ongoing maintenance, indicators of future problems can and do arise. Organizations that formally search for and manage accident precursors can continually find opportunities for improving safety and can thereby reduce the probability of disasters.

REFERENCES

Armstrong, J.S. 1985. Long-Range Forecasting: From Crystal Ball to Computer. New York: John Wiley and Sons.

Asch, D.A., and R.M. Parker. 1988. The Libby Zion case: one step forward or two steps backward? New England Journal of Medicine 318(12): 771–775.

ASRS (Aviation Safety Reporting System). 2001. The Office of the NASA Aviation Safety Reporting System. Callback 260. Moffet Field, Calif.: National Aeronautics and Space Administration.

ASRS. 2003. ASRS Program Overview. Available online: *http://asrs.arc.nasa.gov/overview_nf.htm.*

Barach, P., and S.D. Small. 2000. Reporting and preventing medical mishaps: lessons from non-medical near miss reporting systems. British Medical Journal 320(7237): 759–763.

Bates, D.W., L.L. Leape, D.J. Cullen, N. Laird, L.A. Petersen, J.M. Teich, E. Burdick, M. Hickey, S. Kleefield, B. Shea, M. Vander Vliet, and D.L. Seger. 1998. Effect of computerized physician order entry and a team intervention on prevention of serious medication errors. Journal of the American Medical Association 280(15): 1311–1316.

Bates, D.W., J.M. Teich, J. Lee, D. Seger, G.J. Kuperman, N. Ma'Luf, D. Boyle, and L. Leape. 1999. The impact of computerized physician order entry on medication error prevention. Journal of the American Medical Informatics Association 6(4): 313–321.

Battles, J.B., H.S. Kaplan, T.W. Van der Schaaf, and C.E. Shea. 1998. The attributes of medical event-reporting systems: experience with a prototype medical event-reporting system for transfusion medicine. Archives of Pathology and Laboratory Medicine 122(3): 231–238.

BEA (Bureau d'enquetes et d'analyses pour la securite de l'aviation civile). 2002. Accident on 25 July 2000 at "La Patte d'Oie" in Gonesse (95), to the Concorde, registered F-BTSC operated by Air France. Paris: Ministere de l'equipement des transports et du logement. Available online: *http://www.bea-fr.org/docspa/2000/f-sc000725pa/pdf/f-sc000725pa.pdf.*

Bedford, T., and R. Cooke. 2001. Probabilistic Risk Analysis: Foundations and Methods. Cambridge, U.K.: Cambridge University Press.

Bier, V.M. 1993. Statistical methods for the use of accident precursor data in estimating the frequency of rare events. Reliability Engineering and System Safety 41: 267–280.

Bier, V.M., Y.Y. Haimes, J.H. Lambert, N.C. Matalas, and R. Zimmerman. 1999. A survey of approaches for assessing and managing the risk of extremes. Risk Analysis 19(1): 83–94.

Bier, V.M., and A. Mosleh. 1990. The analysis of accident precursors and near misses: implications for risk assessment and risk management. Reliability Engineering and System Safety 27(1): 91–101.

Bird, F.E., and G.L. Germain. 1996. Practical Loss Control Leadership. Revised ed. Calgary, Alberta: Det Norske Veritas.

Birkmeyer, J.D., C.M. Birkmeyer, D.E. Wennberg, and M.P. Young. 2000. Leapfrog Safety Standards: Potential Benefits of Universal Adoption. Washington, D.C.: The Leapfrog Group.

Bourrier, M. 1996. Organizing maintenance work at two nuclear power plants. Journal of Contingencies and Crisis Management 4: 104–112.

Bridges, W.G. 2000. Get Near Misses Reported, Process Industry Incidents: Investigation Protocols, Case Histories, Lessons Learned. Pp. 379–400 in Proceedings of the International Conference and Workshop on Process Industry Incidents: Investigation Technologies, Case Histories, and Lessons Learned. October 2, 5, 6, 2000. New York: American Institute of Chemical Engineers.

BTS (Bureau of Transportation Statistics). 2002a. Project 6 Overview: Develop Better Data on Accident Precursors or Leading Indicators. In Safety in Numbers Conference Compendium. Washington, D.C.: Bureau of Transportation Statistics.

BTS. 2002b. Project 7 Overview: Expand the Collection of "Near-Miss" Data to All Modes. In Safety in Numbers Conference Compendium. Washington, D.C.: Bureau of Transportation Statistics. Available online: *http://www.bts.gov/publications/safety_in_numbers_conference_2002/project07/project7_overview.html.*

CAIB (Columbia Accident Investigation Board). 2003. Columbia Accident Investigation Board Report. Vol. 1. Washington, D.C.: National Aeronautics and Space Administration. Available online at: *www.caib.us/news/report.*

Carroll, J.S., and S. Hatakenaka. 2001. Driving organizational change in the midst of crisis. MIT Sloan Management Review 42(3): 70–79.

Chess, C., A. Saville, M. Tamuz, and M. Greenberg. 1992. The organizational links between risk communication and risk managment: the case of Sybron Chemicals Inc. Risk Analysis 12(3): 431–438.

Chiles, J.R. 2002. Inviting Disaster: Lessons from the Edge of Technology. New York: HarperCollins.

CIRAS (Confidential Incident Reporting and Analysis System). 2003. CIRAS Executive Report. Glasgow, U.K.: CIRAS.

Classen, D. 2003. Engineering a Safer Medication System Creating a National Standard. Presentation to the National Academy of Engineering/Institute of Medicine Workshop on Engineering and the Health Care System, February 6–7, 2003, Irvine, California.

Cook, R., D. Woods, and C. Miller. 1998. A Tale of Two Stories: Contrasting Views of Patient Safety. Chicago: National Patient Safety Foundation.

Cooke, R., and L. Goossens. 1990. The Accident Sequence Precursor methodology for the European post-Seveso era. Reliability Engineering and System Safety 27: 117–130.

CSB (Chemical Safety Board). 2002. Investigation Report: Chemical Manufacturing Incident. NTIS PB2000-107721. Washington, D.C.: Chemical Safety Board.

Cullen, W.D. 2000. The Ladbroke Grove Rail Inquiry. Norwich, U.K.: Her Majesty's Stationery Office.

DOT-FAA (U.S. Department of Transportation, Federal Aviation Administration). 1997. Advisory Circular Aviation Safety Action Programs (ASAP), AC# 120-66. Washington, D.C.: Federal Aviation Administration.

DOT-FAA. 2000. Advisory Circular Aviation Safety Action Programs (ASAP), AC# 120-66A. Washington, D.C.: Federal Aviation Administration.

DOT-FAA. 2002. Advisory Circular: Aviation Safety Action Programs. AC# 120-66B. Washington, D.C.: U.S. Department of Transportation.

DOT-FAA. 2003. Advisory Circular: Aviation Safety Action Programs. AC# 120-66C. Washington, D.C.: U.S. Department of Transportation.

DOT-NHTSA (U.S. Department of Transportation, National Highway Traffic Safety Administration). 2002. Reporting of Information and Documents About Potential Defects Retention of Records That Could Indicate Defects; Final Rule, CFR, Vol. 67, No. 132. Washington, D.C.: U.S. Department of Transportation.

Dowell, A.M., and D.C. Hendershot. 1997. No good deed goes unpunished: case studies of incidents and potential incidents caused by protective systems. Process Safety Progress 16(3): 132–139.

Er, J., H.C. Kunreuther, and I. Rosenthal. 1998. Utilizing third-party inspections for preventing major chemical accidents. Risk Analysis 18(2): 145–153.

Evans, R.S., S.L. Pestotnik, D.C. Classen, T.P. Clemmer, L.K. Weaver, J.F. Orme, J.F. Lloyd, and J.P. Burke. 1998. A computer assisted management program for antibiotics and other anti-infective agents. New England Journal of Medicine 338(4): 232–238.

Fischhoff, B. 1975. Hindsight = / = foresight: the effect of outcome knowledge on judgment under uncertainty. Journal of Experimental Psychology: Human Perception and Performance 1: 288–299.

Freudenburg, W.R. 1988. Perceived risk, real risk: social science and the art of probabilistic risk assessment. Science 242 (4875): 44–49.

GAO (General Accounting Office). 1998. U.S. efforts to implement flight operational quality assurance programs. Aviation Safety 17(7-9): 1–36.

Hawkins, S.A., and R. Hastie. 1990. Hindsight: biased judgments of past events after the outcomes are known. Psychological Bulletin 107: 311–327.

IOM (Institute of Medicine). 2000. To Err Is Human: Building a Safer Health System, L.T. Kohn, J.M. Corrigan, and M.S. Donaldson, eds. Washington, D.C.: National Academies Press.

Johnson, J.W., and D.M. Rasmuson. 1996. The US NRC's Accident Sequence Precursor Program: an overview and development of a Bayesian approach to estimate core damage frequency using precursor information. Reliability Engineering and System Safety 53: 205–216.

Jones, S., C. Kirchsteiger, and W. Bjerke. 1999. The importance of near miss reporting to further improve safety performance. Journal of Loss Prevention in the Process Industries 12: 59–67.

Kelly, B.D., and M.S. Clancy. 2001. Use a comprehensive database to better manage process safety. Chemical Engineering Progress 97(8): 67–69.

Kilbridge, P., and D. Classen. 2002. Surveillance for Adverse Drug Events: History, Methods and Current Issues. VHA Research Series, Vol. 2. Irving, Texas: Veterans Health Administration.

Kletz, T. 1994. Learning from Accidents, 2nd ed. Oxford, U.K.: Butterworth-Heinemann.

Kumamoto, H., and E.J. Henley. 2000. Probabilistic Risk Assessment and Management for Engineers and Scientists. New York: John Wiley and Sons.

Kunreuther, H.C., P.J. McNulty, and Y. Kang. 2002. Third-party inspection as an alternative to command and control regulation. Risk Analysis 22(2): 309–318.

Kunreuther, H.C., S. Metzenbaum, and P. Schmeidler. 2003. Leveraging the Private Sector: Management-Based Strategies for Improving Environmental Performance. Paper Presented at Conference on Leveraging the Private Sector: Management-Based Strategies for Improving Environmental Performance, July 31–August 1, 2003, Resources for the Future, Washington, D.C.

Lakats, L.M., and M.E. Paté-Cornell. In press. Organizational warning systems: a probabilistic approach to optimal design. IEEE Transactions on Engineering Management 51(2).

LaPorte, T.R. 1988. The United States Air Traffic System: Increasing Reliability in the Midst of Rapid Growth. Pp. 215–244 in The Development of Large-Scale Technical Systems, R. Mayntz and T. Hughes, eds. Boulder, Colo.: Westview Press.

LaPorte, T.R., and P. Consolini. 1998. Theoretical and operational challenges of "high-reliability organizations": air-traffic control and aircraft carriers. International Journal of Public Administration 21: 847–852.

Leapfrog Group. 2003. The Leapfrog Group Factsheet: Computerized Physician Order Entry System. Revision 4/18/03. Washington, D.C.: The Leapfrog Group.

Lodwick, D.G. 1993. Rocky Flats and the evolution of distrust. Research in Social Problems and Public Policy 5: 149–170.

March, J.G., L.S. Sproull, and M. Tamuz. 1991. Learning from samples of one or fewer. Organization Science 2(1): 1–13.

Marcus, A.A. 1995. Managing with danger. Industrial and Environmental Crisis Quarterly 9(2): 139–152.

Marcus, A.A., and M.L. Nichols. 1999. On the edge: heeding the warnings of unusual events. Organization Science 10(4): 482–499.

Martin, B. 1999. Suppression of dissent in science. Research in Social Problems and Public Policy 7: 105–135.

Minarick, J.W., and C.A. Kukielka. 1982. Precursors to Potential Severe Core Damage Accidents: 1969–1979, A Status Report. NUREG/CR-2497. Washington, D.C.: U.S. Nuclear Regulatory Commission.

O'Neill, A.C., L.A. Petersen, E.F. Cook, D.W. Bates, T.H. Lee, and T.A. Brennan. 1993. Physician reporting compared with medical-record review to identify adverse medical events. Annals of Internal Medicine 119(5): 370–376.

Paté-Cornell, M.E. 1986. Warning systems in risk management. Risk Analysis 5(2): 223–234.

Paté-Cornell, M.E., and P. Fischbeck. 1993. Probabilistic risk analysis and risk-based priority scale for the tiles of the space shuttle. Reliability Engineering and System Safety 40(3): 221–238.

Paté-Cornell, M.E., and S.D. Guikema. 2002. Probabilistic modeling of terrorist threats: a systems analysis approach to setting priorities among countermeasures. Military Operations Research 7(4): 5–23.

Phimister, J.R., U. Oktem, P.R. Kleindorfer, and H. Kunreuther. 2003. Near miss incident management in the chemical process industry. Risk Analysis 23(3): 445–459.

Pidgeon, N.F. 1991. Safety culture and risk management in organizations. Work and Stress 12(3): 202–216.

Pooley, E. 1996. Nuclear warriors. *Time*, March 4, pp. 46–54.

Reisch, F. 1994. The IAEA asset approach to avoiding accidents is to recognize the precursors to prevent incidents. Nuclear Safety 35: 25–35.

Roberts, K.H. 1990. Some characteristics of one type of high reliability organization. Organization Science 1(2): 160–176.

Rochlin, G.I. 1999. Safe operation as a social construct. Ergonomics 42(3): 1–12.

Rochlin, G.I., T.R. LaPorte, and K.H. Roberts. 1987. The self-designing high-reliability organization: aircraft carrier flight operations at sea. Naval War College Review 40(4): 76–90.

Strauss, B., and M.G. Morgan. 2002. Everyday threats to aircraft safety. Issues in Science and Technology 19(2): 82–86.

Turner, B.M., and N. Pidgeon. 1997. Man-made Disasters, 2nd ed. London: Butterworth-Heinemann.

USNRC (U.S. Nuclear Regulatory Commission). 1978. Risk Assessment Review Group Report. NUREG/CR-0400. Washington, D.C.: Nuclear Regulatory Commission.

van der Schaaf, T.W. 1992. Near Miss Reporting in the Chemical Process Industry. Ph.D. Thesis, Eindhoven University of Technology, the Netherlands

van der Shaff, T.W., D.A. Lucas, and A.R. Hale, eds. 1991. Near Miss Reporting as a Safety Tool. Oxford, U.K.: Butterworth-Heineman.

Vaughan, D. 1997. The Challenger Launch Decision: Risky Technology, Culture, and Deviance at NASA. Chicago: University of Chicago Press.

Volpe Center. 2003. Improving Safety through Understanding Close Calls. Cambridge, Mass.: Volpe Center.

Weick, K.E., and K.H. Roberts. 1993. Collective mind and organizational reliability: the case of flight operations on an aircraft carrier deck. Administrative Science Quarterly 38: 357–381.

Weick, K.E., and K.M. Sutcliffe. 2001. Managing the Unexpected: Assuring High Performance in an Age of Complexity, Vol. 1. New York: John Wiley and Sons.

Weick, K.E., K.M. Sutcliffe, and D. Obstfeld. 1999. Organizing for high reliability. Pp. 81–123 in Research in Organization Behavior 21, R.S. Sutton and B.M. Staw, eds. Stamford, Conn.: JAI Press.

Westrum, R., and A.J. Adamski. 1999. Organizational factors associated with safety and mission success in aviation environments. Pp. 67–104 in Handbook of Aviation Human Factors, D.J. Garland, J.A. Wise, and V.D. Hopkin, eds. Mahwah, N.J.: Lawrence Erlbaum Associates.

Wiegmann, D.A., H. Zhang, T. von Thaden, G. Sharma, and A. Mitchell. 2002. A Synthesis of Safety Culture and Safety Climate Research. Technical Report ARL-02-3/FAA-02-2. Urbana-Champagne, Ill.: Aviation Research Laboratory, Institute of Aviation, University of Illinois.

White House. 2000. President Clinton Announces New Public-Private Partnerships to Increase Aviation Safety. Press release, January 14. Washington, D.C.: Office of the Press Secretary, White House.

Section II
Keynote Speakers

The Opportunity of Precursors

JAMES P. BAGIAN
U.S. Department of Veterans Affairs
National Center for Patient Safety

One difficulty in identifying vulnerabilities in a system, sometimes called the precursor problem, is hindsight bias. After a big, smoking hole appears in the ground, it is very easy to say someone should have taken the problem seriously. That bias certainly appeared in the wake of the Three Mile Island incident and the *Challenger* and *Columbia* space shuttle disasters. But, as people with operational or hands-on managerial experience know, in any large, complex project, people often bare their souls and express their uncertainties about many aspects of a project at the last minute. Often these last minute revelations are attempts to prevent being held responsible for a bad outcome—in the case of the space shuttle, the deaths of seven people who were strapped in and launched on that day. The manager, at whose desk the buck finally stops, has to ask what data support these last minute concerns.

Even if the data are not very good, decisions must be made. Concerns about possible negative outcomes, although they must be taken into account, should not inordinately influence a final decision, which should be based on facts and not emotions. Every project entails risks, which can never be eliminated entirely. Nevertheless, when a bad outcome occurs, the knee-jerk response is to equate it with a bad decision. When the causes have been analyzed, however, they may very well show that the decisions leading up to the bad outcome were entirely appropriate.

The real challenge we face is how to go from theory to practice. In making that practical transition, it is essential that we first determine the ultimate goal. Unless a goal is clearly understood—not in tactical terms but in terms of the end result—confusion and ineffective actions are likely to result. For instance, we might ask what the ultimate goal of manufacturing buggy whips was—to make transportation using horses more efficient or to enable people, merchandise, and information to move over large distances as quickly as possible. If we understand that the latter was the goal, then clearly other modalities, such as cars, trucks, airplanes, etc., should have been pursued as they became available. The changeover to more effective modalities is not always instantaneous though, because it is easy to become enamored of a particular technique, hobby, or traditional way of achieving a goal. So we must always ask ourselves what the overarching goal is and how we can best reach it.

Systems in the health-care field are not always clearly aligned with the goals of the overall system. If we ask about the overarching goal of patient safety initiatives, the answer is usually to eliminate errors. The problem with this answer is that eliminating errors is a tactic, not a goal. Of course, it is impossible to eliminate all errors. Therefore, adopting a goal of eliminating all errors is tantamount to declaring that a project is doomed to fail. It may sound simplistic, but failing to distinguish between goals and tactics can result in efforts that do not lead to solutions of the problems at hand. Activities should be measured against the yardstick of whether or not they really contribute to meeting strategic objectives.

In medicine, our goal should be to prevent unnecessary harm to the patient, not to eliminate errors. People involved in health care may disagree about standards of care or about what constitutes an error. But when patients end up injured or dead, and these outcomes were avoidable, everyone agrees that these outcomes are a bad thing.

RECOGNIZING OUTCOMES

The aftermath of the Three Mile Island incident is instructive for understanding the perception of outcomes. Two divergent views of the incident emerged that were polar opposites and raised questions about how we define success. One side fretted that the accident demonstrated that nuclear power was dangerous. People who drew this conclusion tended to view any risk to life as undesirable or unacceptable. The other side felt that the plant's safety systems had achieved their goals by preventing a disaster. Both views are worthy of consideration, but in fairness, the yardstick for evaluation must be to measure performance against design specifications. If, in hindsight, the design specifications are determined to be inadequate, then the specifications should be revised, but the performance of the system on the day in question is not the primary issue.

The traditional approach to recognizing a problem is reactive and retrospective, which appears to be a natural human response. Unfortunately, the perceptions created by this less-than-scientific, or hindsight-biased, approach can unduly influence behavior. September 11, 2001, for instance, made a huge impression on the way people think about terrorism. After that day, many people refused to fly on airplanes—some still refuse today. Arguably, flying is no riskier today than it was before September 11, but people perceive it differently now, and those perceptions govern their decisions.

Similar rethinking about aviation has happened before. Until the late 1940s, airplane crashes were in many ways regarded as "the cost of doing business." Statistics from World War II show that the number of planes lost as a result of normal, noncombat activities was staggering compared to the number lost to enemy fire. At the time, this was not regarded as abnormal, but as the way of the world. Some would call the acceptance of the risk of airplane crashes the "normalization of deviance."

The attitude toward plane crashes began to change in the 1950s. Accident investigation data from the U.S. Navy (similar to data from other military services) show that the aircraft loss rate has dramatically dropped since then. In 1950, there were approximately 54 losses per 100,000 flying hours; 776 aircraft were lost in 1954 alone. By 1996, the figure had dropped to approximately two aircraft per 100,000 flying hours. That's a 96 percent reduction, even though the physical environment in which pilots operated (i.e., low-level, high-speed, all-weather, and night flights) presented many more objective hazards than before. This reduction was accomplished through the institution of a proactive and systematic approach to safety.

REPORTING SYSTEMS

We study precursors because we want to take a new, proactive approach to system safety that emphasizes prevention. To become proactive, however, we must first identify problems that could lead to bad outcomes. One of the tools for becoming proactive is an effective reporting system. But because there can be tremendous barriers to reporting, it is essential that the ultimate goals of the reporting system be clearly defined. One of the first decisions that must be made is whether the purpose of reporting is organizational learning or accountability. Safety systems that have a goal of preventing harm in a sustainable way must emphasize organizational learning.

Another important decision is what should be reported. Is the purpose of the reporting system to look only at things that have caused an undesirable outcome, or is it also to scrutinize other things, such as close calls that almost resulted in undesirable outcomes but did not, either because of a last-minute "good catch" or because of plain good fortune? Close calls are extremely important areas of study because they are much more common than actual bad outcomes. Thus, close calls provide repeated opportunities to learn without first having to experience a tragic outcome. In addition, because close calls do not result in harm, people are generally more willing to discuss them openly and candidly, because they are less fearful of retribution for the part they played in the event. Also, people are often more motivated to analyze close calls if understanding them is considered an opportunity to act proactively to prevent undesirable outcomes in the future.

However, in medicine as in industry, close calls are often ignored. For instance, the Joint Commission on Accreditation of Healthcare Organizations, a quasi-regulatory body, recognizes the value of analyzing close calls but does not require that they be investigated. In fact, if someone submits a voluntary report of a close call with an associated root-cause analysis (RCA) to the Joint Commission, it will not be accepted. This policy sends a mixed message that appears to emphasize learning only from events in which patients have been injured rather than from close calls where learning can take place without first having to hurt a patient.

Great care must be taken in using data in reporting systems. By their very nature, the self-reporting that populates most reporting systems cannot be used to estimate the true incidence of events. This fact is often overlooked, as has been demonstrated by erroneous incidence statistics published after analysts have "tortured the data" from a variety of reporting systems. We must remember that reports are simply reports; they do not necessarily reflect reality. Trends and rates based on them are simply trends and rates of what was reported, which may bear no relation to the trends and rates of the entire system. The best use of reporting systems is for identifying potential system vulnerabilities.

We had a report, for example, that identified a significant problem. A patient was physically pinned to a magnetic resonance imaging (MRI) scanner by the "sandbags" that had been used to stabilize him. MRI scanners have very strong magnets, and sandbags are sometimes inappropriately filled with ferromagnetic particles rather than sand. Had we relied on the so-called rate and incidence statistics culled from our reporting system, we would have concluded that this was not an important problem, because we had never received previous reports of such problems in MRI suites. However, we thought this represented a real vulnerability, so we went out to medical centers, both inside and outside the VA, to observe, talk to people, and learn what was happening on a daily basis. We found that similar system issues were quite common. For example, MRIs often caused pens and paper clips to fly out of shirt pockets, sometimes striking patients.

As a result of this fieldwork, we implemented a system-wide alert with instructions for mitigating these risks to our patients. If we had relied on misleading statistics based on reports, we would have ignored the single report and decided that it was not worth studying the problem.

A reporting system is essential, but it is also essential not to become a slave to it. Reporting systems and self-reports never totally reflect reality, but they are valuable for identifying system vulnerabilities. But if we only sit in our offices counting and sorting reports, it is unlikely that anything will get better.

ANALYZING PRECURSORS

Reports can be thought of as fuel for the safety-improvement engine. How then do we get this fuel? A mandatory reporting system is not the answer. Although it may seem like a simple solution, it ignores real issues concerning the effective relationship that must be developed with those from whom you wish to receive reports. Anyone who thinks people will report adverse events just because there is a regulation that they do so is living in a dream world. As Dr. Charles Billings, the father of the Aviation Safety Report System stated, "in the final analysis all reporting is voluntary" (statement at a meeting of Advisory Panel on Patient Safety Systems, Washington, D.C., March 12, 1998). In other words, there is no such thing as mandatory reporting. Billings says people only report what they care to report, either because there are penalties for not reporting and the event was witnessed by others or because they feel there is some intrinsic value to reporting an event to improve the system. People will not report simply because there is rule that they do so. Senate Bill 720 and House Bill 663 recognize the fallacy of mandatory reporting requirements and endorse voluntary reporting.

There are numerous sources of information about hazards and risks. The challenge becomes determining how to prioritize reports and what to do with the information. The VA has developed a prioritization methodology based on the severity of an event and its probability of occurrence; we assign each event a safety assessment code, which determines if a detailed RCA is required.

In determining action to be taken, it is essential to look at the root causes and contributing factors that led to an undesirable condition or event. There is seldom a single cause. A thorough analysis of underlying causes can provide insight into the problem and a basis for taking steps to correct or prevent the problem. For instance, we looked at a collection of RCAs dealing with cases in which incorrect surgical procedures were performed or incorrect sites were operated on. The RCAs revealed that the problem was much different than had been thought.

It had been generally accepted that the problem was mainly one of identifying the correct side of the body to operate on. After looking at the real situation, we found that marking to establish laterality was an issue, but almost as big an issue was that the incorrect patient was operated on because of inadequate patient identification methods. In many cases, the incorrect patient was scheduled to undergo a similar procedure the same day by the same physician, but on another part (or side) of the body. Only by understanding the underlying causes could we take appropriate countermeasures and implement preventive strategies.

CREATING LEARNING SYSTEMS

There are many accountability systems in medicine but very few learning systems. Most medical problems and errors that occur today have happened before and will continue to happen in the future unless we do something differently. It is naïve to expect that the traditional approach of punishing the individual(s) involved in an incident will make the world safer. Rather than assigning blame, we must start learning from mistakes and translating the lessons into system-based solutions.

One of the most important steps in creating a learning system is demonstrating to participants in the system that the objective is not punishment but systematic improvements that will prevent undesirable events from occurring in the future. For a learning system to be trusted, it must be considered fair, in plan and in practice. This does not mean that it must be a blame-free system.

We have found that frontline staff do not want a blame-free system; they want a fair system. Therefore, we established a ground rule that the results of an RCA could not be used to punish an individual. However, because we did not want the safety system to be used, or appear to be used, to hide events that all parties agree require disciplinary action, we decided to define events that were "blameworthy." We did not use legal terms, such as reckless or careless, that have been interpreted differently in different jurisdictions. Instead, we defined a blameworthy act as an "intentionally unsafe act," that is, a criminal act, an act committed under the influence of drugs or alcohol, or a purposefully unsafe act.

Blameworthy acts are not examined as part of the safety system, which is strictly for learning. They are passed on to the administrative system where, besides learning, punishment may be an outcome; in addition, the proceedings are also discoverable. Thus, those who wish to report events clearly understand under what circumstances they can be subject to punitive action. Although I cannot prove causality, I can state that, after we instituted these definitions, reporting went up 30-fold in an organization that already had a good reporting culture.

A learning system must be shown to be useful. People in the organization must understand why the system is necessary and what its benefits might be. People will not waste time reporting if their participation makes no difference. If the system becomes a black hole into which learning and energy disappear, people will not participate.

The individual is the most important component of an effective learning system. The system is absolutely necessary to making things work, but it ultimately depends on people. Systems that work best are the ones people internalize as their own. A successful system creates an environment that makes people want to do the right thing. In systems that work, people speak up and communicate with each other.

In a study last year, pilots and board-certified physicians were asked if they would protest if their superiors told them to do something with which they disagreed. Virtually all of the pilots said yes, but only about half of the physicians did. This shows vastly different thresholds for communicating critical information.

TAKING THE LEAD

This workshop is not about management; it is about leadership. Successful leaders must be willing to take on risk. One example of leaders assuming risk is the way the VA issues patient safety alerts. Alerts identify a discrete problem, describe its solution, and set a time by which the solution must be implemented. We identify so many problems through our reporting system that it would be easy to issue 100 alerts a day. But we know that issuing too many alerts would ultimately make people indifferent to alerts, thus creating new risks. Therefore, we prioritize potential alerts through a scoring mechanism, and we issue only the most critical alerts for national dissemination.

So far, this approach has resulted in an average of three or four alerts per month. We realize that this approach could put leadership in a politically awkward position someday if a vulnerability for which we chose not to issue an alert resulted in a patient injury. But rather than taking a self-serving, risk-averse position and sending out many more alerts, thus passing the responsibility and risk to the front line, VA leadership believes we can more effectively help patients by issuing alerts judiciously, even though the leadership is placed in greater personal/professional jeopardy. Leaders should be willing to accept the risk of being criticized in exchange for a safer system for patients.

Leaders must demonstrate priorities by their actions, as well as their words. Paying them lip service is not enough. The old aphorism, lead by example, is still true. Leaders must maintain a relentless drumbeat that safety-related activities are an inextricable part of everything we do.

There is altruism out there, and people will participate in a reporting system that they feel is fair and that provides a safe environment for them. People cannot be forced to participate; they must be invited to play. In the VA patient safety program, we have demonstrated that the best way of getting people to participate initially is to make program adoption and implementation voluntary. In a pilot test, this approach attracted dedicated volunteers, and within a few weeks, the response to the program was so favorable that the remaining facilities asked to implement the program. We completed the pilot test and rolled the program out nationally in just nine months. In fact, facilities that had initially been reluctant to adopt the new system became impatient when they were told they would have to wait their turn to be trained in the use of the new system. It is possible to win enthusiastic acceptance, but this requires patience and not trying to force acceptance.

SUMMARY

The purpose of a reporting system is to identify vulnerabilities; the desired result should be preventing harm to patients. Once the causes of underlying vulnerabilities have been determined, corrective action can be implemented. System improvements should be made based on the study of causes and vulnerabilities. However, if the reporting system does not result in actions that mitigate the vulnerabilities locally or throughout a system, then the entire effort is for naught.

Perhaps the most important resource for learning is close calls. Under the old VA reporting system, reports of close calls accounted for only 0.05 percent of all reports. In the current system, which emphasizes close calls, 95 percent of reports are about close calls. Another important breakthrough was ensuring that events that were reported resulted in action being taken. Under the old system, less than 50 percent of all events that received in-depth analysis resulted in any action being taken. This created great cynicism. With the new system, less than 1 percent of RCAs do not result in corrective action(s). Not every action is effective, of course, but if no action is taken, it is certain that the situation will not improve.

We typically use a different team for every RCA. In this way, we provide experiential learning for staff members, who come to appreciate the value and details of the safety system. When they return to their jobs, their view of the world is very different. The response to this changing-team approach has been almost uniformly positive.

In the end, success is not about counting reports. It is about identifying vulnerabilities and precursors to problems and then formulating and implementing corrective actions. Analysis and action are the keys, and success is manifested by changes in the culture of the workplace. Change does not happen overnight; it takes time.

As Einstein said, "The significant problems we face cannot be solved with the same level of thinking we were at when we created them." His corollary to this was that "Insanity is doing the same thing over and over and expecting different results." But probably most important of all for us today is something Margaret Mead said, "Never doubt that a small group of thoughtful, committed people can change the world. Indeed it's the only thing that ever has."

On Signals, Response, and Risk Mitigation

A Probabilistic Approach to the Detection and Analysis of Precursors

ELISABETH PATÉ-CORNELL
Department of Management Science and Engineering
Stanford University

Precursors provide invaluable signals that action has to be taken, sometimes quickly, to prevent an accident. A probabilistic risk analysis coupled with a measure of the quality of the signal (rates of false positives and false negatives) can be a powerful tool for identifying and interpreting meaningful information, provided that an organization is equipped to do so, appropriate channels have been established for accurate communication, and mechanisms are in place for filtering information and reacting to true alerts (Paté-Cornell, 1986). The objective is to find and fix system weaknesses and reduce the risks of failure as much as possible within resource constraints (Paté-Cornell, 2002a).

In this paper, I present several examples of failures and successes in the monitoring of system performance and an analytical framework for optimizing warning thresholds. I then briefly discuss the application of this reasoning to the modeling of terrorist threats and the characteristics of effective organizational warning systems.

THE SPACE SHUTTLE

In 1988, in the wake of the *Challenger* disaster, the National Aeronautics and Space Administration (NASA) offered me the opportunity to study, as part of my research, some of the subsystems of the space shuttle. Conversations with the head of the mission-safety office at NASA headquarters and with some of the astronauts revealed that they were concerned about the tiles that protected the shuttle from excessive heat during reentry. Therefore, I decided that the tiles

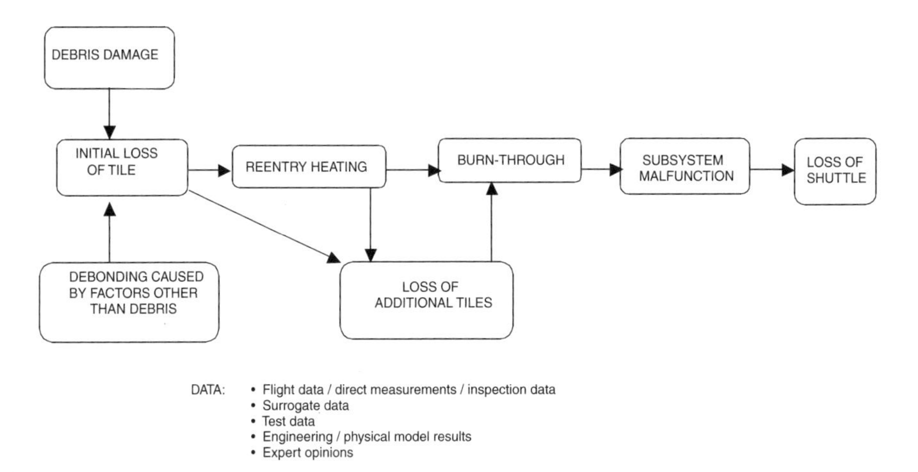

FIGURE 1 Influence diagram for an analysis of the risk of an accident caused by the failure of tiles on the space shuttle. Source: Paté-Cornell and Fischbeck, 1994.

were an appropriate subject for a risk analysis that might reveal some fundamental problems and help avert an accident.

With NASA funding, and with the assistance of one of my graduate students (Paul Fischbeck), I performed such an analysis based on the first 33 flights of the shuttle. The results were published in several places (Paté-Cornell and Fischbeck, 1990, 1993a,b, 1994). I went first to Johnson Space Center (JSC) to get a better understanding of how the tiles worked, what problems might arise, and how the tiles might fail. The study was based on four critical parameters for each tile: (1) the heat load, which is vitally important because, if a tile is lost, the aluminum skin at that location might melt, thus exposing the internal systems to hot gases; (2) aerodynamic forces because, if a tile is lost, the resultant cavity creates a turbulence that could cause the next tile to fail; (3) the density of hits by debris, which might indicate the vulnerability of the tile to this kind of load; and finally (4) the criticality of the subsystems under the skin to determine the consequences of a "burn-through" in various locations of the orbiter's surface. Based on these four factors, we constructed a risk analysis model (Figure 1) described as an influence diagram.

The pattern of debris hits was intriguing. First, we looked at maps of direct hits the shuttle had experienced during each of its 33 flights (a map of hits had been recorded for each flight). When we superimposed these maps, we found an interesting pattern of damage under the right wing (Paté-Cornell and Fischbeck, 1993a,b). As it turns out, a fuel line runs along the external tank on the right side, and because of the way the foam insulation on the external tank is applied, little pieces of insulation had debonded where the fuel line was attached to the tank.

This observation immediately directed our attention to what was happening with the insulation of the external tank, as well as the system's performance under regular loads (e.g., vibrations, aerodynamic forces, etc.).

The next question we examined was what the consequences would be if the aluminum skin were pierced in different locations of the orbiter's surface. We found that once a tile or several tiles was lost, the aluminum skin would be exposed; it would begin to soften at approximately 700°C and would melt shortly above that temperature. In some places, a burn-through would be catastrophic. For example, the loss of the hydraulic lines or the avionics, would lead to an accident.

Once it was clear that the tile system was critical, I wanted to understand the factors that affected the capacity of the tiles to withstand the different loads to which they were subjected. I went to Kennedy Space Center (KSC) to talk to the tile technicians and observe their work. In the course of these discussions, I discovered that during maintenance, a few tiles had been found to be very poorly bonded. This could have happened, for example, if the glue had been allowed to dry before pressure was applied, either during the first installation or later during maintenance. Even though poorly bonded tiles could withstand the 10-pounds-per-square-inch pull test, they could be dislodged either by a large debris hit or, perhaps, even by normal loads, such as high levels of vibration. At JSC, I also asked for the potential trajectories of debris that could debond from the insulation of the external tank, both from the top and the center of the tank (Paté-Cornell and Fischbeck, 1990, 1993b). At Mach 1, it seemed that tiles debonded from either location would hit the tiles under the wings. These trajectories appear in the original report (Paté-Cornell and Fischbeck, 1990). I must point out, however, that in general I did not look into the reinforced carbon-carbon, including on the edge of the left wing, which seems to have been hit first in the *Columbia* accident of February 2003.

In December 1990, I delivered a report to NASA pointing out serious problems, both with the foam on the external tank and the weak bonding of some of the tiles (Paté-Cornell and Fischbeck, 1990). One of the findings was that about 15 percent of the tiles were responsible for 85 percent of the probability per flight of a loss of vehicle and crew due to a failure of the tiles. The risk of an accident caused by the tiles was evaluated at that time to be approximately 1/1,000 per flight.

Next, we constructed a map of the underside of the orbiter to show the location of the most risk-critical tiles, so that when NASA (as required by the procedures in place) picked 10 percent of the tiles for detailed inspection before each flight, the technicians would have an idea of where to begin (Figure 2).

But obviously, most of the risk was the result of the potential for human error, in many cases a direct consequence of management decisions. Therefore, I also looked into some management issues. I learned that tile technicians were paid a bit less than machinists and other technicians, so they tended to move on

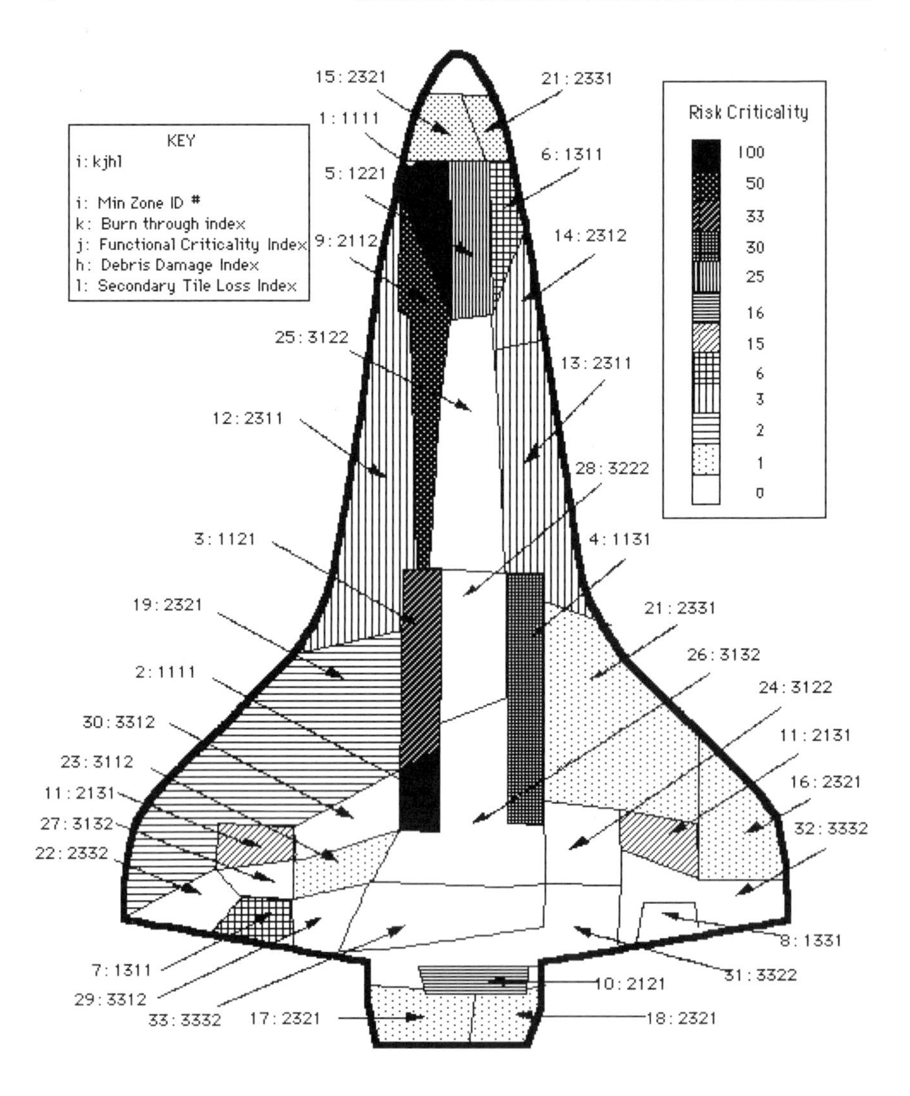

FIGURE 2 Map of the risk criticality of the tiles on the space shuttle orbiter as a function of their location. Source: Paté-Cornell, 1990, 1993a,b, 1994.

to other jobs. Therefore, the tile maintenance crews sometimes lost some experienced workers. I also learned that tile technicians at the time were under considerable pressure to finish work on the spacecraft quickly for the next flight. Because of those time constraints, some workers had become creative—for instance, at least one of them had decided to spit into the tile glue to make it cure faster. But the curing of the glue is a catalytic reaction and adding water to the bond at the time of curing could perhaps cause it to revert to a liquid state sooner than it would otherwise.

The completed study was published in the literature (Paté-Cornell and Fischbeck, 1993a,b). In 1994, we were among the finalists for the Edelman prize of the Institute for Operations Research and Management Sciences (INFORMS) for that work (Paté-Cornell and Fischbeck, 1994). We were told by the jury that we were not chosen because we could not "prove" that if NASA implemented our recommendations, it would save the agency some money. That proof, unfortunately, came with the *Columbia* accident.

Shortly thereafter, the study was revived by Dr. Joseph Fragola, vice president and principal scientist at Science Applications International Corporation (SAIC), who incorporated it into a complete risk analysis of the shuttle orbiter. After that, it seems that the study was essentially forgotten, except for efforts at JSC in recent years to revisit it to try to lower the calculated risks of an accident caused by a tile failure. In any case, NASA lost the report, and, with some embarrassment, asked me for a copy of it on February 2, 2003.

On the morning of the accident (February 1, 2003), I was awakened by a phone call from press services asking for my opinion about what had just happened. I did not immediately conclude that a piece of debris that had struck the left wing at takeoff had been the only cause of the accident as described in one of the scenarios of the 1990 report. But I knew immediately that it could not possibly have helped for the shuttle to have reentered the atmosphere with a gap in the heat shield.

Had NASA implemented the study's recommendations? In fact, quite a few of the problems noted in the 1990 report about organizational matters had been corrected. For instance, the wages of tile technicians had been raised, eliminating some of the turnover among those workers, and the risk-criticality map had been used at KSC to prioritize tile inspections. But it appears that at JSC, where maintenance procedures are set, management had concluded that the study did not justify modifying current procedures. As a result, unfortunately, several things that should have been done were not. For example, no nondestructive methods were effectively developed for testing the tile bond. Tests could have been done using ultrasounds, which would have been expensive but, with sufficient resources, might have been achieved by now. Second, once in orbit, the astronauts were unable to fix gaps in the heat shield. Imagine that you are in orbit looking down to reentry and you realize one or more tiles are missing. To me, this was a real nightmare. At the present time (after the accident), NASA seems to have concluded that the astronauts should have the skills to fix tiles in flight before the space shuttles fly again. That process may be completed as early as December 2003.

NASA might also have looked at the precursors, especially the poorly bonded tiles, and done something about them. Instead, in its risk analyses, NASA *redefined* the precursors. The 1990 study had concluded that 10 percent of the risk of a shuttle accident could be attributed to the tiles. But apparently, NASA thought this figure was too high because a number of flights had occurred without any tile loss since our study. So they asked a contractor to redo the analysis; the contractor decided to take as a precursor the number of tiles lost in flight (instead of the number of weakly bonded tiles). During the first 68 flights, only one tile had been lost, some of the felt had remained in the cavity, and the lost tile had not caused an accident. Obviously, the new analysis changed the results, and the computed risk went down from 1/1,000 to 1/3,000.

I believe that the contractor focused on the wrong precursor, that is, a phenomenon (the number of tiles that had debonded in flight) for which statistics were insufficient. Indeed, history corrected the new results when two additional tiles were subsequently lost, which brought the risk result back to about 1/1,000. Therefore, I believe that our original study had used a better precursor, because it provided sufficient evidence to show that the capacity of a number of tiles was reduced before they actually debonded or an orbiter was lost.

FORD-FIRESTONE

The Ford-Firestone fiasco is another example of precursors being ignored until it was too late. The Firestone tires on Ford Explorer SUVs blew out at a surprisingly high rate, causing accidents that were sometimes deadly. But it took 500 injuries and roughly 150 deaths before Ford reacted. The first signals had been detected by State Farm Insurance in 1998, but nothing was done about the problem in the United States.

Part of the problem was a split warranty system. The car was under warranty by Ford, but the tires were under warranty by Firestone. This created a data-filtering problem that has now been addressed by the TREAD Act passed by Congress in 2000. The TREAD Act mandates the creation of an early warning system; the law requires that even minor problems be reported to the National Highway Safety Transportation Administration (DOT-NHTSA, 2001). Some serious questions, however, have not yet been fully addressed—what data should be collected, how data should be stored, and how the information should be organized and processed. My research team is currently studying these issues.

SUCCESS STORIES

In spite of some visible failures, there have been some real successes in the creation of systems designed to identify and observe precursors. William Runciman, then head of anesthesiology of the medical school in Adelaide, Australia, constructed the Australian incident monitoring study (AIMS) database (Webb et al., 1993). The system requires that every anesthesiologist submit an anonymous report after emerging from the operating room describing problems encountered during the operation. Because anonymity is absolute, the questionnaire is widely

TABLE 1 Incidence Rates of Initiating Events in Anesthesia Accidents (gathered from the AIM database)

Initiating Event	Number of AIMS Reports*	Report Rate	Probability of Initiating Event	Relative Fraction
Breathing circuit disconnect	80	10%	7.2×10^{-4}	34%
Esophageal intubation	29	10%	2.6×10^{-4}	12%
Nonventilation	90	10%	8.1×10^{-4}	38%
Malignant hyperthermia	n/a	—	1.3×10^{-5}	1%
Anesthetic overdose	20	10%	1.8×10^{-4}	8%
Anaphylactic reaction	27	20%	1.2×10^{-4}	6%
Severe hemorrhage	n/a	—	2.5×10^{-5}	1%

*Out of 1,000 total reports in initial AIMS data.

used. Indeed, for some, the exercise seems to be almost cathartic. The AIMS reporting system enables the hospital to identify the frequency and probability of various factors that initiate accident sequences that might kill healthy patients under anesthesia (e.g., people undergoing knee surgery). These factors include, for example, incorrect intubation (e.g., in the stomach instead of the lungs) and overdoses of a particular anesthetic.

A risk analysis we did with a research group from Stanford based on the AIMS database provided insights into improving safety in anesthesia (Paté-Cornell, 1999; Paté-Cornell et al., 1996). Table 1 shows the kind of data we derived from the AIMS database regarding the probability of an initiating event per operation. Figure 3 shows the structure of the risk analysis model in which these data were used to identify (1) the main accident sequences (and their dynamics), (2) the effect of the "state of the anesthesiologist" (e.g., extreme fatigue) on patient risk, and (3) the effect of management procedures (e.g., restriction on the length of time on duty) on the state of the anesthesiologist.

Another success story is airplane maintenance. Some of my students worked, under my supervision, on a project to analyze maintenance data for one of the most popular airplanes in the fleet of a major airline (Sachon and Paté-Cornell, 2000). Their study revealed an intriguing problem. Because of rare errors during maintenance, the flaps and slats of the leading edge of the plane sometimes dropped on one side during flight. There has never been an accident involving these flaps and slats, apparently because the drop never occurred at a critical time. We computed the probability of the drop happening at takeoff, landing, or in bad weather, in other words, the probability of an accident that has not happened yet. Based on our work, the airline decided to modify its maintenance procedure. We hope this work will make a difference in the long run.

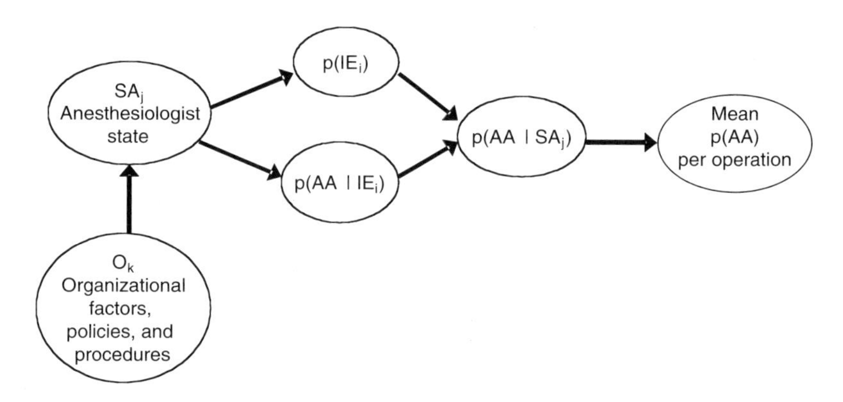

FIGURE 3 Influence diagram showing the analysis of patient risk in anesthesia linked to human and organizational factors. Source: Paté-Cornell et al., 1997.

COMBATING TERRORISM

The U.S. intelligence community failed to recognize precursors leading up to the catastrophic terrorist attack of September 11, 2001. This was a very complex situation. Obviously, signals were missed, or were not allowed to surface, within an agency; but there were also legal issues that prevented early actions. There are two distinct problems, both related to the fusion of information: (1) the combination of different pieces and different kinds of evidence; and (2) communication among agencies.

Before September 11, the U.S. intelligence community was aware of the activities of Al-Qaeda but did not recognize (or act upon) precursors to the attack for many reasons. One aspect of the overall failure was the lack of communication among intelligence agencies. This was partly the result of laws passed at the end of the Vietnam War that mandated separate databases for separate entities, which deliberately kept agencies from communicating with each other. To overcome this problem, we need interfaces between some computers and databases. Congress, however, is reluctant to permit the development of such interfaces for a sound reason—the need to protect privacy. One project by the Defense Advanced Research Projects Agency to implement interfaces was terminated because, although it was legal, some members of Congress thought it was getting too close to an invasion of privacy.

Questions of gathering and processing intelligence are particularly interesting. Two kinds of uncertainty problems are involved: (1) a statistical problem of extracting relevant information from background noise; and (2) the problem of identifying and gathering information about possible threats that have been detected but have not been confirmed. The first problem can be enormously difficult, like finding a needle in a haystack hidden by an opponent intent on deception. But even if we have several pieces of intelligence information, some strong and some weak, some independent and some not, the challenge is to determine the probability of an attack of a given type in the next specified time window given what we know today.

Bayesian reasoning involving base rates, as well as the likelihood of the signals given the event, can be extremely helpful in addressing this challenge. Bayesian reasoning allows the computing of probabilities in the absence of a large statistical database; it uses logical reasoning based on the prior probability of an event and on the probabilities of errors (both false positives and false negatives) (Paté-Cornell, 2002a).

I began doing some risk modeling of terrorism as a member of a small panel of the Air Force Scientific Advisory Board. The panel members included eminent specialists in many fields, from history of the Arab world to weapons design. The topic under study was asymmetric warfare. We had to deal with a great mass of information under a relatively tight deadline. To help set priorities, I constructed for myself a probabilistic risk analysis model in the form of an influence diagram (Figure 4).

For simplicity, I considered only two kinds of terrorist groups—Islamic fundamentalists and disgruntled Americans (Paté-Cornell and Guikema, 2002). Experts then told me that although these groups had different preferences, they had some common characteristics, such as the importance of the symbolism of the target. I then looked more carefully at what we know about the supply chain of different terrorist groups (people and their skills, materials, transportation, communications, and cash). It was clear that both the supply chains and the preferences of the American groups were different from those of Islamist groups. Next, I asked how likely a group was to have U.S. insiders assisting them. I was then able to put the characteristics of an attack scenario into three classes: (1) choice of weapons; (2) choice of targets; and (3) means of delivery. For instance, the weapon might be smallpox or a nuclear warhead. I also considered the possibility of repeated urban attacks.

With the assistance of one of my graduate students, Seth Guikema, I constructed a "game" model involving two diagrams, one for the terrorists and one for the U.S. government (Paté-Cornell and Guikema, 2002). Based on our beliefs about the terrorists' knowledge and preferences and our perceptions of the preferences of the American people in general, and the administration in particular, we then looked for feedback between the two models to try to set priorities among attack scenarios.

Obviously, a large part of the relevant information is classified, and the published paper contains only illustrative numbers. But, as expected, it appears that one attractive weapon for Islamist groups is likely to be a nuclear warhead—if they can get their hands on one—because of the sheer destruction it would cause. Smallpox was second in the *illustrative* ranking. Dirty bombs (e.g., spent

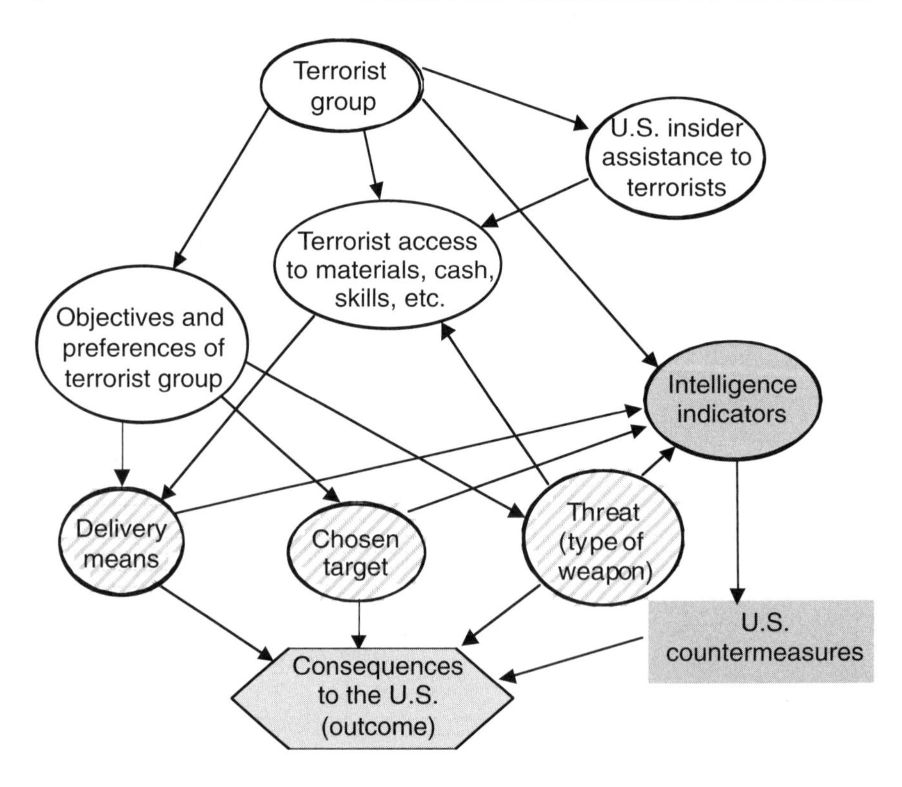

FIGURE 4 Influence diagram representing an overarching model for prioritizing threats and countermeasures. Source: Paté-Cornell and Guikema, 2002.

fuel combined with conventional explosives) were lower on the scale because, although they are scary and easy to make, they may not do as much damage.

The U.S. Department of Homeland Security's recent simulation exercise (TOPOFF 2) was also instructive. TOPOFF 2 involved a hypothetical combined attack on Seattle with a dirty bomb and on Chicago O'Hare Airport with biological weapons. One problem faced by the participants was uncertainty about the plume from the dirty bomb. The models constructed by the Nuclear Regulatory Commission and the Environmental Protection Agency, perhaps because they were created for regulatory purposes, turned out to be too conservative to be very helpful in predicting the most likely shape of the plume. Those models are thus unlikely to match actual measurements, which may undermine confidence in the analytical results.

In any case, modeling attack scenarios has to be a dynamic exercise. Technologies change every day, and terrorist groups are constantly evolving, which, of course, makes long-term planning difficult. Therefore, it may be most useful to start by identifying desirable targets. For example, infrastructure is important to the average person, but perhaps not very attractive to terrorists because it lacks symbolic impact. Before these groups strike at infrastructure targets, they may try, given the opportunity, to hit something that looks more appealing in that respect.

So what can we do? Of course, we should identify the weak points in global infrastructure systems (e.g., the electric grid), because they are worth reinforcing in any case. We should also make protecting symbolic buildings, prominent people, harbors, and borders, a high priority.

Monitoring the supply chains of potential terrorist groups (people, their skills, the materials they use, cash, transportation [both of people and materials], and communications) is especially important. A coherent and systematic analysis of the intelligence signals remains difficult and often depends on intuition, but a Bayesian analysis of the precursors and signals would account explicitly for reliability and dependencies (Paté-Cornell, 2002a). But we have a long way to go before a probabilistic analysis will be implemented, because it is not part of a tradition. Currently some interest has been expressed in using such methods, and I hope that this logical, organized approach will eventually be adopted.

ORGANIZATIONAL WARNING SYSTEMS

An effective organizational warning system has to be embedded in an organization's structure, procedures, and culture. The purpose of a warning system is to watch for potential problems in organizations that construct, operate, or manage complex, critical physical systems.

How do we begin to think about an organizational warning system? First, we must analyze, in parallel, the dynamics of the physical system and of the organization. As Figure 5 shows, a good place to start is by identifying the *physical* weaknesses of a critical system, using a probabilistic risk analysis, for example, as a basis for setting priorities (look at the engines before you look at the coffee pot). The second step is to find corresponding signals and decide what the monitoring priorities should be. The third step is to examine the dynamics of both the organization and the system as it evolves. Once a problem starts, how rapidly does a situation deteriorate, and how soon does a failure occur if something is not done? Finally, part of the analysis is to assess the error rate of the monitoring system to provide filters and to prevent counterproductive clogging of the communication channels.

The next step is to analyze the organizational processing of the message. How is an observed signal communicated? Who is notified? Who can keep the signal from being transmitted? Does the decision maker eventually hear the message (and how soon), or is it lost—or the messenger suppressed—along the way? Finally, how long does it take for a response to be actually implemented? Clearly, no response can be formulated without an explicit value system for weighing

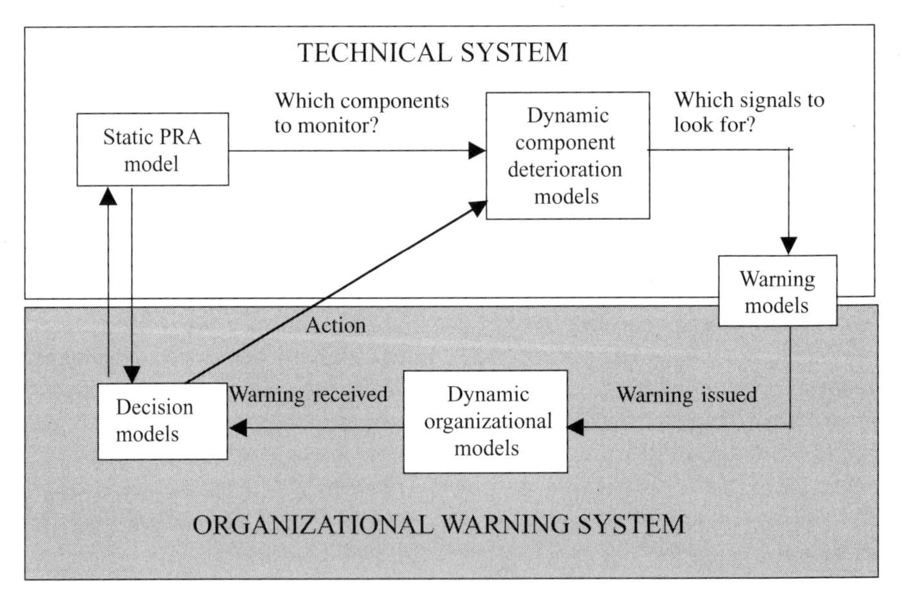

FIGURE 5 Elements of a warning system management model. Source: Lakats and Paté-Cornell, 2004.

costs and benefits, as well as the trade-offs between false positives and false negatives.

SETTING AN ALERT THRESHOLD

One way of filtering the signal is to optimize the warning threshold based on an explicit stochastic process (Figure 6) that represents the variations over time of the underlying phenomenon to be monitored (Paté-Cornell, 1986). The problem could be, for instance, the density of smoke in a living space as determined by a fire alarm. How sensitive should the system be? If it is too sensitive, people may shut it off because of too many false alerts. If it is not sensitive enough, the situation can deteriorate too much before action is taken. The warning threshold should thus be a level at which people will respond and that gives them enough lead time to react to the signal.

An appropriate alarm level is likely to lead to some false alerts, which have costs. For instance, a false positive indicating a major terrorist attack could lead to costly human risks—for instance, the risks incurred in a mass evacuation. A false positive could also cause a loss of confidence that reduces the response to the next alert. False negatives (missed signals) also imply some costs. If no warning is issued when it should be, there is nothing one can do, and losses will be incurred. Another cost of a false negative is loss of trust in the system. To

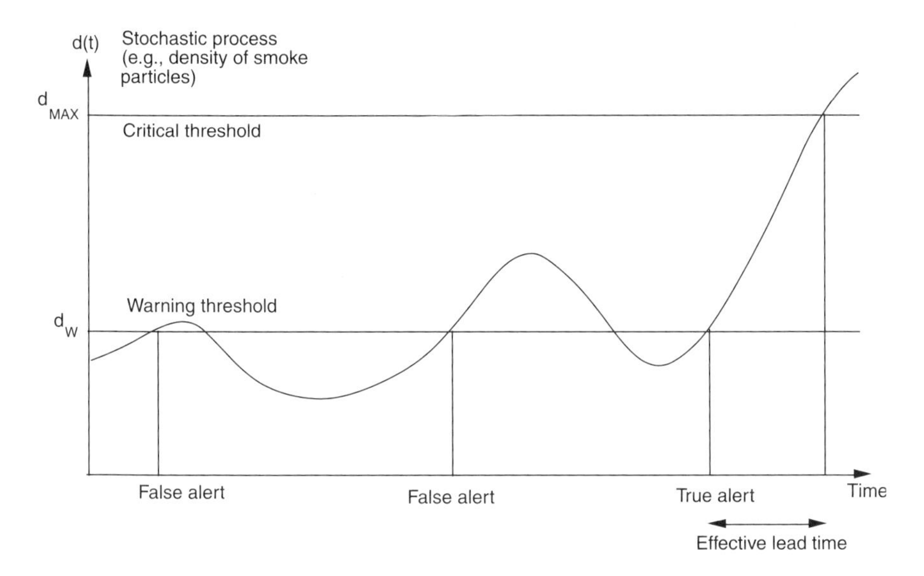

FIGURE 6 Optimization of the threshold of a warning system based on variations of the underlying stochastic process. Source: Paté-Cornell, 1986.

provide a global assessment of a warning system, the results thus have to be examined case by case, but the base rates and the rates of errors must be taken into account.

Given the trade-offs between false positives and false negatives, there is no way to resolve the problem of filtering out undesirable signals without making a value judgment. A balance has to be found between the time necessary for an appropriate response, the corresponding benefits, and the cost of false alerts, in terms of both money and human reaction. Therefore, a simple model of this process must include at least three elements: (1) the underlying phenomenon (in terms of recurrence and consequences); (2) the rate of response given people's experience with the system's history of true and false alerts; and (3) the effectiveness of the use of lead time.

One of the problems that has emerged with the color alert system used by the U.S. Department of Homeland Security is a lack of clarity about what an alert means and what should be done. For the time being, however, it is not clear how the system can be improved. Some have suggested using a numerical scale, for instance, but this would have to be very well thought through to be more meaningful than the color alert system. In any case, determining the cost and benefits of a warning system involves taking into account the base rate and the effects of the events, as well as false positives and false negatives.

SUMMARY

In conclusion, I would like to make several points. First, failure stories do not provide a complete picture. There are many more precursors and signals observed and acted upon than there are accidents. This is not always apparent, however, because when signals are recognized and timely actions taken, the incident is rarely visible, especially, for example, in the domain of intelligence. Therefore, we know when we have failed, but we often do not know when we have succeeded. Second, managing the trade-off between false positives and false negatives in warning systems is difficult because it involves the quality of the information, as well as costs and values. Third, to design an effective organizational warning system, one has to know where to look.

The way people in an organization react depends in part on management. The structure, procedures, and culture of the organization determine the information and the incentives people perceive, as well as the range of their possible responses. It is essential to decide how to filter a message to avoid passing along information that may have many costs and few benefits. In that respect, the quality of communications is essential—both communication to decision makers who must have the relevant facts in hand and communication to the public who must have trust in the system and respond appropriately. And finally, when it comes to risk management, the story is always the same—when it works, no one hears about it...

REFERENCES

DOT-NHTSA (U.S. Department of Transportation-National Highway Transportation Safety Administration). 2001. NPRM: Reporting of Information and Documentation about Potential Defects. Retention of Records that Could Indicate Defects. Pp. 66,190–66,226 in Federal Register 66 (246) Dec. 21, 2001.

Lakats L.M., and M.E. Paté-Cornell. 2004. Organizational warnings and system safety: a probabilistic analysis. IEEE Transactions on Engineering Management 51(2).

Paté-Cornell, M.E. 1986. Warning systems in risk management. Risk Analysis 5(2): 223–234.

Paté-Cornell, M.E. 1999. Medical application of engineering risk analysis and anesthesia patient risk illustration. American Journal of Therapeutics 6(5): 245–255.

Paté-Cornell, M.E. 2002a. Finding and fixing systems weaknesses: probabilistic methods and applications of engineering risk analysis. Risk Analysis 22(2): 319–334.

Paté-Cornell, M.E. 2002b. Fusion of intelligence information: a Bayesian approach. Risk Analysis 22(3): 445–454.

Paté-Cornell, M.E., and P.S. Fischbeck. 1990. Safety of the Thermal Protection System of the STS Orbiter: Quantitative Analysis and Organizational Factors. Phase 1: The Probabilistic Risk Analysis Model and Preliminary Observations. Research report to NASA, Kennedy Space Center. Washington, D.C.: NASA.

Paté-Cornell, M.E., and P.S. Fischbeck. 1993a. Probabilistic risk analysis and risk-based priority scale for the tiles of the space shuttle. Reliability Engineering and System Safety 40(3): 221–238.

Paté-Cornell, M.E., and P.S. Fischbeck. 1993b. PRA as a management tool: organizational factors and risk-based priorities for the maintenance of the tiles of the space shuttle orbiter. Reliability Engineering and System Safety 40(3): 239–257.

Paté-Cornell, M.E., and P.S. Fischbeck. 1994. Risk management for the tiles of the space shuttle. Interfaces 24: 64–86.

Paté-Cornell, M.E., and S.D. Guikema. 2002. Probabilistic modeling of terrorist threats: a systems analysis approach to setting priorities among countermeasures. Military Operations Research 7(4): 5–23.

Paté-Cornell M.E., L.M. Lakats, D.M. Murphy, and D.M. Gaba. 1996. Anesthesia patient risk: a quantitative approach to organizational factors and risk management options. Risk Analysis 17(4): 511–523.

Paté-Cornell, M.E., D.M. Murphy, L.M. Lakats and D. M. Gaba. 1997. Patient risk in anesthesia: probabilistic risk analysis, management effects and improvements. Annals of Operations Research 67(2): 211–233.

Sachon, M., and M.E. Paté-Cornell. 2000. Delays and safety in airline maintenance. Reliability Engineering and Systems Safety 67: 301–309.

Webb, R.K., M. Currie, C.A. Morgan, J.A. Williamson, P. Mackay, W.J. Russel, and W.B. Runciman. 1993. The Australian Incident Monitoring Study: an analysis of 2000 incident reports. Anaesthesia and Intensive Care 21: 520–528.

Section III
Risk Assessment

Understanding Accident Precursors

MICHAL TAMUZ
Health Science Center
University of Tennessee

Organizations seek to identify the factors that might cause or contribute to adverse events before these precursors result in accidents. But understanding accident precursors poses difficulties for organizations that seek to untangle those factors from the snarled mesh of history. An organization attempts to learn from its history of accidents, but these infrequent adverse events yield sparse data from which to draw conclusions (March et al., 1991). The unacceptable cost of such events precludes the usual organizational methods of learning from experience by trial and error (La Porte, 1982). In addition, processes of detecting danger signals can be clouded by ambiguities and uncertainties (Marcus and Nichols, 1999; Weick, 1995) and obscured by redundant layers of protection (Sagan, 1993). Finally, when things go wrong, organizations often use the same data as a basis for disciplining those involved and for identifying accident precursors. But linking data collection with disciplinary enforcement inadvertently creates disincentives for the disclosure of information (Tamuz, 2001). Despite these difficulties, or perhaps because of them, various industries have developed alternative models for detecting and identifying accident precursors.

The types of accidents or adverse events vary among industries, from airplane crashes in the aviation industry to lethal emissions in the chemical industry to patient injuries and deaths in the health care industry. These harmful events differ in their probability of occurrence; in the distribution of their negative consequences among employees, clients, and the public; in the complexity and interdependence embedded in their technologies; and in the regulatory context in which they operate (Perrow, 1984). For example, although the estimated number of deaths and injuries attributed to preventable adverse events in health care far outnumbers the average loss of life in aircraft accidents, the deaths in health care occur one patient at a time and usually without media attention. Aircraft accidents kill many people in one disastrous, highly publicized moment. Indeed, aviation professionals may also lose their lives in a crash (Thomas and Helmreich, 2002). The nuclear power industry, unlike aviation and health care, has to contend with a hostile public skeptical about its ability to operate safely. Organizations in various industries also differ in their capacity to intervene and avert catastrophe (Perrow, 1984; van der Schaaf, p. 119 in this volume).

Despite these and other critical differences, decision makers in diverse industries are all engaged in a common search for accident precursors. Some industries, such as aviation and nuclear power, have a relatively long history of seeking to identify accident precursors; others, such as blood banks and hospitals, are relative newcomers to the field. Nevertheless, they all use similar information-gathering processes and weigh common design choices. Whereas some industries discovered precursors based on their common experiences, such as having to draw on small samples of accidents (March et al., 1991), other industries developed precursor detection programs as a result of learning by imitation (Levitt and March, 1988), such as in the Patient Safety Reporting System.

SEEKING ACCIDENT PRECURSORS AMONG NEAR ACCIDENTS

Accidents and adverse events provide critical sources of information about accident precursors. Discovering precursors from accidents, however, can be difficult, because accidents can be infrequent, costly, and complex (Tamuz, 1987). In industries such as nuclear power and aviation where accidents are rare, organizations investigate accidents in great detail, but they have few accidents from which to learn. In sorting through the complex circumstances of a single accident, it may be difficult to ascertain whether specific conditions preceding the accident are precursors or just a coincidence of random events. Furthermore, adverse events have the potential for catastrophic consequences, not only for those directly involved, but also for the organizations involved—the hospital, airline, device manufacturer, or blood bank—which may be held liable for damages. The actual contributing factors to the accident may be obscured in the struggle to establish liability by casting blame (Tasca, 1989).

Near accidents, events in which no damages or injuries occur but, under slightly different circumstances, could have resulted in harm, are important sources of information about accident precursors (NRC, 1980; Tamuz, 1987). Methods of gathering and sorting near accident data to reveal precursors have been developed in high-hazard industries, in which accidents are rare but have disastrous consequences. The air transportation industry builds on a common experiential understanding of a near miss, such as when two aircraft nearly collide. The nuclear power industry and chemical manufacturing industry draw on engineering culture, in which it has long been assumed that only chance differentiates near accidents from industrial accidents (Heinrich, 1931).

Terms for a near accident, such as a "close call" and a "near miss," are being adapted by health care organizations. This reflects a change in emphasis from assessing actual harm to patients, as expressed by the traditional admonition to do no harm, to evaluating the potential for adverse outcomes (Stalhandske et al., 2002). Adopting lessons from the aviation industry, hospital transfusion-medicine departments (Battles et al., 1998), the Department of Veterans Affairs hospital system (Heget et al., 2002), and the Agency for Healthcare Research and Quality (e.g., Pace et al., 2003) have promoted the implementation of close-call reporting systems.

To provide an overview for discussing accident precursors, this paper is divided into two sections. First, the Aviation Safety Reporting System (ASRS) is described to illustrate some common processes involved in detecting and identifying accident precursors and to provide a common frame of reference. A basic understanding of how ASRS identifies accident precursors is important not only for understanding aviation safety programs, but also because it has become a widely discussed and adapted model in health care (IOM, 2000; Leape, 1994). Second, based on examples in the aviation, nuclear power, and health care industries, a few key design choices and trade-offs are described.

PROCESSES IN IDENTIFYING ACCIDENT PRECURSORS

Sifting through the shards of near accidents, organizations engage in several processes to identify precursors. Building on a model developed in a previous accident precursor workshop (Brannigan et al., 1998), I propose that these processes include: aggregating data, detecting signals, gathering information, interpreting and analyzing information, making and implementing decisions, compiling and storing data, and disseminating information.

Although the processes are listed in order, they often occur in recurring decision-making loops linked by feedback chains. For example, in the process of analyzing an event, safety analysts may decide to gather additional information about procedures that preceded the event. Returning to the information-gathering process, they search their database for reports of similar procedures. The location of such feedback loops can be essential for promoting (or impeding) learning. For example, hospital pharmacists diligently gathered data about the prescribing errors they had prevented by calling and asking for clarifications from the resident physician who ordered the medication. But the pharmacy did not provide feedback to the residents and those who train them; thus, the lack of a feedback loop linking the pharmacy back to the physicians may have hindered efforts to identify precursors to adverse drug events (Tamuz et al., 2004).

In practice, organizations skip some processes. To identify threats, decision makers may rely on data collected for regulatory purposes rather than gathering data solely for the purpose of detecting precursors. For example, data collected in air traffic control centers to monitor controllers' operational errors and pilots' deviations from regulations have also been used to identify hazardous conditions (Tamuz, 2001). Similarly, when ASRS identifies reports that illustrate well known, albeit sometimes overlooked, precursors, it proceeds directly from the processes of interpretation and analysis of information to the dissemination of educational information.

ASRS is a voluntary, confidential, nonpunitive reporting system, managed under the auspices of the National Aeronautics and Space Administration, funded by the Federal Aviation Administration (FAA), and operated by a long-term contractor. The following brief description of ASRS is based on interviews with key participants, supporting documents provided by them (Reynard et al., 1986), and secondary sources (Connell, p. 139 in this volume; National Academy of Public Administration, 1994).

The following discussion describes a conceptual model constructed from the processes of identifying accident precursors applied to ASRS operations (summarized in Table 1).

Aggregating Data

Data regarding safety-related events are compiled at a national level. Individuals working in airlines, airports, and air traffic control facilities are encouraged to file reports. Although pilots submit most of the reports, ASRS encourages reporting by air traffic controllers and other groups in the aviation community.

Detecting Signals

The goal of ASRS is to detect signals of potentially dangerous events by having individuals in the air transportation industry report their perceptions of safety-related incidents. The definition of an event is broadly defined, and individuals are encouraged to report anything they perceive to be significant. In practice, as will be explained below, pilots have incentives to report events that could involve violations of FAA regulations. ASRS specifically excludes reports of accidents, events that resulted in injuries or property damage, and intentional regulatory violations, such as sabotage.

Gathering Information

ASRS contributes to the identification of accident precursors by collecting data in ways that overcome some of the traditional barriers to gathering reports of negative information in organizations. First, reporting is voluntary. Second, it is promoted as a professional obligation. Indeed, instruction in the use of ASRS

TABLE 1 Processes in Identifying Accident Precursors

	Aviation Safety Reporting System
Aggregating data	Data on a national level
Detecting signals	Safety-related incidents, excluding accidents, criminal acts, and intentional regulatory violations
Gathering information	Voluntary, professional reporting system Incentives, including limited immunity from prosecution for pilots Confidentiality, including call-back capacity and de-identification
Interpreting and analyzing data	Classification of events for safety significance Identification of urgent hazardous situations Focus on potential outcomes (what could have occurred) Identification of examples of known precursors Prospective view: discovery of possible accident precursors for further investigation Retrospective view: investigation of accidents in context of near accidents
Compiling and storing data	Centralized compilation Public data distribution
Making and implementing decisions	Recommendations to FAA Lack of decision-making authority
Disseminating information	Distribution of hazard warnings Provision of data to regulators and public Publication of practical precursor information

reporting forms is routinely included in the training of general aviation pilots, and the Airline Pilots Association encourages its members to file ASRS reports. Third, pilots receive limited immunity from administrative action by the FAA if they have filed an ASRS report regarding a possible infraction. The FAA may still take action against the pilot, but the sanctions will not include certificate suspension, thus allowing the pilot to continue flying. Finally, ASRS uses de-identification of both individuals and airlines to ensure that reports are not used as a basis for regulatory enforcement, disciplinary action, or litigation. The main objective of information gathering is to promote learning from experience rather than to discipline individuals for regulatory infractions.

In response to its innovative data gathering methods, ASRS receives different kinds of reports. Some pilots submit sparse, telegraph-style factual summaries, describing their aircraft's deviation from an assigned altitude, for example. These reports appear to be filed simply to protect the pilot from possible FAA enforcement action. Other reports of events that involve regulatory violations describe in detail how conditions that caused near accidents could have resulted in accidents. A third type of report describes potentially dangerous events, with no mention of a possible air traffic violation. Thus, although ASRS provides incentives for pilots to file reports, some pilots take the time to report events they perceive to be dangerous, even if they do not benefit directly from the incentives (Tamuz, 1987).

Interpreting and Analyzing Information

ASRS analysts first classify reports by their safety significance. If a report has safety potential, it is carefully examined and coded. Potentially significant events are further classified as (1) urgent situations that require immediate intervention or (2) events that warrant in-depth analysis and coding by ASRS safety analysts (Tamuz, 2000). Although ASRS safety analysts occasionally "call back" individuals to obtain additional information, ASRS does not independently investigate reports.

ASRS safety analysts identify accident precursors by examining critical incidents in detail and by noticing patterns in the data. After an aircraft accident, ASRS safety analysts routinely search the database for near accidents that occurred under similar conditions. After identifying a critical near accident, they scan the database to generate hypotheses about potential accident precursors, inform the FAA, and call for further study and, possibly, corrective measures. ASRS analysts also conduct database searches in response to FAA inquiries about potential threats to safety. In addition, ASRS analysts look for instructive examples to illustrate recognized precursors for publications and training materials.

To discern possible accident precursors, ASRS relies mainly on human expertise. ASRS safety analysts are drawn from the ranks of retired pilots and air traffic controllers who have years of experience, as well as expertise. Building on their cumulative knowledge and experience, ASRS has developed an extensive coding scheme for classifying incident reports. Because ASRS is a voluntary reporting system, however, fluctuations in the number of reports are not reliable indicators of changes in underlying safety conditions.

Making and Implementing Decisions

ASRS representatives advise the FAA and make policy recommendations, but they cannot initiate changes. ASRS is designed to gather reports of potential dangers and identify possible accident precursors, but it does not have the authority to make or implement decisions.

Analyses of ASRS near-accident data have led to the identification of many accident precursors. For example, ASRS analysts noticed that skilled pilots had almost lost control of their aircraft in the wake turbulence from a Boeing 757 aircraft, even though they were following at the prescribed distance (wake turbulence can be described as horizontal, tornado-like vortices of air behind an aircraft). Tragically, these ASRS data did not reach the appropriate FAA decision makers, and no corrective action was taken until several accidents attributed to wake turbulence had occurred (Reynard, 1994). This illustrates ASRS's capacity to identify accident precursors proactively, as well as the importance of feedback loops linking ASRS data analysts to FAA policy makers.

Compiling and Storing Data

The ASRS data compilation activities are performed under carefully defined confidentiality restrictions. Indeed, analyses and conclusions drawn by safety analysts are not released to the public. ASRS staff do conduct some limited searches of the database for the aviation community, researchers, and the public. De-identified ASRS data are also available on the Web, as part of the FAA National Aviation Safety Data Analysis Center, and are routinely used by journalists.

Disseminating Information

ASRS representatives regularly brief FAA policy makers and issue warnings about hazardous situations in air traffic control facilities and airports. ASRS also publishes information about accident precursors on the Web and in print. In particular, they disseminate information about accident precursors and other threats to safety to individuals working in aviation (Hardy, 1990). These examples are published in "Callback," a newsletter distributed to members of the aviation community and freely available on the Web. For example, an issue of "Callback" featured the following statement in an excerpt from a pilot's ASRS report, "As I was accelerating down the runway, a shadow appeared." The shadow was of another aircraft landing immediately in front of him (Callback, 2003). The editors used this example to call attention to a well known, but overlooked accident precursor.

KEY DESIGN CHOICES AND TRADE-OFFS

Industries, and the organizations within them, differ in their methods of identifying accident precursors. They may differ in their design choices for the level of aggregating data, in how they define and classify safety-related events, in their choice of surveillance or reporting systems, and in their methods of overcoming barriers to reporting. Each of these choices can result in trade-offs that influence the system's capacity for identifying accident precursors.

Aggregating Data: Pooling Data on an Organizational or Interorganizational Level

Aviation

The implications of aggregating data at the organizational or interorganizational level are apparent in a comparison of an airline-based model, the Airline Safety Action Partnership (ASAP), with the nationwide model, ASRS. ASAP was created when representatives of the Southwest Region of the FAA Flight Standards Division joined with the pilots' association and management of American Airlines to promote the confidential disclosure and correction of potentially dangerous conditions, ranging from inadequate techniques demonstrated by an individual pilot to the identification of accident precursors (Aviation Daily, 1996). In an airline-based reporting system, safety analysts not only investigate organizational conditions to determine if they constitute accident precursors, but they also have the expertise to identify necessary changes and the decision-making authority to eliminate precursor conditions or mitigate their effects.

For example, ASAP members work with (1) union representatives on the ASAP committee to urge pilots to take remedial training; (2) management representatives to change airline procedures; and (3) FAA committee members to influence regulatory changes. Indeed, based on ASAP reports, the airline has identified and used accident precursors in pilot training sessions, updated unclear and potentially confusing airline procedures, and clarified regulatory expectations.

By comparison, ASRS aggregates data at a national level, which enables safety analysts to identify patterns in rare events that, if reported only to an airline, might be classified as isolated events. However, the de-identification of airlines that enables ASRS to gather reports from pilots from competing airlines impedes the gathering of data on specific airline operations. Thus, decision makers cannot detect and correct airline-specific precursors based on ASRS data.

Aggregating data at an organizational or interorganizational level is presented here as a design choice. In some situations, however, the choices (and the trade-offs) need not be made. The British Airways Safety Information System (BASIS) was originally designed as an airline-based system (Holtom, 1991). However, with the widespread adoption of the BASIS model by other airlines, the system was expanded to enable the pooling of data among airlines. Thus, BASIS members benefit from the advantages of pooling data on an interorganizational level and the capacity for corrective action of an airline-based system.

Nuclear Power

The advantage of pooling data at the national level is apparent in the Accident Sequence Precursors Program, a national system sponsored by the U.S.

Nuclear Regulatory Commission (USNRC) for gathering and analyzing data from the required licensee event reports (LERs) of serious near accidents at nuclear power plants (Minarick, 1990; Sattison, p. 89 in this volume). Because significant events, such as LERs, occur infrequently at any one plant, data from all relevant nuclear power plants must be aggregated. If, however, potentially dangerous events occur that do not meet the LER definitions, they are not reported to the USNRC. Hence the trade-off. Although events classified as nonreportable by LER criteria may provide information about accident precursors, the data remain within the particular plant and would be aggregated at the plant level, rather than the national level, although the USNRC representative located at the plant may also know about the nonreportable events. In addition, the Institute of Nuclear Power Operations (INPO) encourages, but does not require, plant managers to report such events to a closed information dissemination system operated by INPO.

Health Care

Health care organizations, including blood banks, hospitals, and pharmacies, have begun to pool data regarding actual and potential adverse events. The Medical Event Reporting System for Transfusion Medicine (MERS-TM) maintains a database that enables transfusion medicine departments at participating hospitals to pool data on errors and near misses (Battles et al., 1998). Each department has access to its own data and the pooled data, but no department has access to another department's specific data. In a similar arrangement, the Veterans Health Administration (VHA) Patient Safety Reporting System collects data on close calls and patient safety issues from individuals working in VHA hospitals across the country. VHA also continues to support hospital-based, close-call reporting systems (Bagian, p. 37 in this volume; Heget et al., 2002).

The Institute for Safe Medication Practices (ISMP) has established an interorganizational clearinghouse for data from pharmacies and pharmacists on potential and actual adverse drug events, such as mix-ups resulting from drugs with sound-alike names or look-alike packaging (Cohen, 1999). ISMP publishes a newsletter and disseminates warning notices to the professional pharmacy community, mainly lessons learned from the experience of others.

All of these health care organizations have initiated programs for pooling data and have established databases to encourage the identification of precursors. As these innovative programs develop, individual health care organizations may be able to identify precursors from pooled data.

Detecting Signals: Defining and Classifying Safety-Related Events

Aviation

The method used to classify safety-related events can influence an organization's capacity to identify accident precursors. One design choice is between broad or precise definitions. By using a broad, general definition of safety-related events, ASRS can capture data on events that may lead to the identification of previously unknown accident precursors. By contrast, the FAA-operated computerized surveillance system used in air traffic control centers applied specific, precisely measured definitions of deviations from safety standards that tended to identify well known conditions that were unlikely to yield new insights into accident precursors (Tamuz, 2001).

Potentially dangerous events may be "defined away" if the conditions do not meet the technical definition of a safety-related event. An example of defining away a potential danger is a near miss over La Guardia Airport that air traffic controllers did not report because, technically, it did not fit the formal definition of an operational error (Tamuz, 2000). Although in this case the controller could not be held accountable for making an error, the near miss represented a significant threat to safety and was a possible source of precursor information.

The classification scheme also influences an organization's capacity to gather and analyze data about potentially dangerous events. If a safety-related event is classified as an error or a regulatory violation, it can lead to measures designed to maintain individual or organizational accountability. In air traffic control centers, for example, when two aircraft failed to maintain the prescribed distance between them, the event could alternatively be defined as an "operational error" for controllers or a "pilot deviation," depending on who was held accountable. These similar events with differing labels were analyzed and stored in separate databases, hindering the search for possible common precursors (Tamuz, 2001).

Health Care

The classification of safety-related events not only influences how these events are detected, but also enhances (or constrains) an organization's capacity to investigate and draw conclusions from its experience. In an Australian hospital, for example, nurses interpreted and defined away potentially harmful events (Baker, 1997); and in one U.S. hospital pharmacy, the definition of a reportable error led to under-reporting and reduced the flow of medication error data to the hospital, while simultaneously enabling learning within the pharmacy department (Tamuz et al., 2004).

By contrast, in a blood bank, the detection of safety-related events that could not harm patients, but were nonetheless classified as posing a threat to the organization (e.g., prompting a regulatory inspection) triggered the allocation of organizational resources for investigation and problem solving (Tamuz et al., 2001). Hence, these studies of health care organizations suggest that the definition of safety-related events and their classification into alternative categories influence event detection, as well as the activation of organizational routines for gathering and analyzing information.

Gathering Information: Surveillance vs. Reporting Systems

Surveillance and reporting systems are alternative methods of monitoring known accident precursors and discovering new ones. Data about threats to safety can be gathered either through surveillance (i.e., direct observation or auditing) or through voluntary reporting systems.

Aviation

Automated safety surveillance systems have been implemented in the air transportation industry. As early as 1986, the FAA implemented a computerized surveillance system in air traffic control centers that automatically detected when an aircraft failed to maintain its assigned separation distance (Tamuz, 1987). Since then, United Airlines has championed the Flight Operational Quality Assurance Program (other airlines support similar programs) based on technologies for monitoring aircraft operations by collecting real-time flight data, such as engine temperature and flight trajectory (Flight Safety Digest, 1998).

One critical trade-off between using a surveillance system and using a reporting system is between data reliability and the richness of information. In automated surveillance systems, counts of safety-related events, such as tallies of operational errors in air traffic control, tend to be more reliable than data obtained through reporting systems. The number of safety-related events submitted to reporting systems, for example, fluctuates with changes in perceived incentives for reporting (Tamuz, 1987, 2001). Computerized surveillance systems provide reliable monitoring of operational errors and adverse events; however, they may be less useful in detecting the contributing factors that lead to a malfunction or harmful outcome.

The trade-off between data reliability and information richness is illustrated by voluntary reporting systems, such as ASRS and ASAP. The data gathered by these reporting systems do not provide reliable indicators of the frequency of safety-related accidents, but they do enable the identification of accident precursors. Analyses of ASRS reports, for example, have revealed conditions that contribute to accidents, from the well documented consequences of failing to follow standardized landing procedures to the seemingly trivial, but potentially lethal distraction of drinking coffee in the cockpit, which resulted in the enactment of regulations for a sterile cockpit. In the highly publicized case of the B757 wake turbulence, ASRS reports revealed that skilled pilots almost lost control of their aircraft even when they maintained the prescribed distance for aircraft trailing a B757 (Reynard, 1994). Hence, although voluntary reporting systems cannot reliably monitor the frequency of errors and adverse events, they can provide important data that may reveal previously overlooked or unknown precursors.

Health Care

Although individual hospitals have long maintained reporting systems, the underreporting of errors and adverse events is widespread (e.g., IOM, 2000). Adverse drug events in hospitals, for example, are routinely underreported (e.g., Cullen et al., 1995; Edlavitch, 1988). Underreporting to hospital incident reporting systems has been attributed to many factors including shared perceptions of team members (Edmondson, 1996), fear of punishment, and lack of time (Vincent et al., 1999). One design alternative, as noted previously, is to implement close call reporting systems; another is to rely on surveillance rather than on reporting.

Surveillance methods used in hospitals range from traditional labor-intensive methods to new computerized surveillance systems. In some hospitals, nurses and medical researchers periodically audit patient charts to identify errors and adverse events after they have occurred. Other hospitals use sophisticated information technology to identify preventable medical injuries, such as adverse drug events. One of the advantages of these automated surveillance systems is that they can provide a more accurate count of adverse events than reporting systems (Bates et al., 2003).

Gathering Information: Overcoming Barriers to Reporting

Aviation

A regulatory agency must make a critical trade-off between its responsibility to maintain accountability and its responsibility to identify and avert accident precursors (Tamuz, 2001). This is reflected in the necessity of choosing between engaging in regulatory enforcement and foregoing punishment to encourage event reporting. Consider the design choices in the FAA Near Midair Collision Reporting System and ASRS. If a pilot reports a near miss to the FAA Near Midair Collision Reporting System in which an air traffic regulation was violated, the FAA can initiate enforcement action against the pilot based on his own report. If the pilot reports the same near midair collision to ASRS, he may be eligible for immunity. Thus, the design of the FAA Near Midair Collision Reporting System creates disincentives for reporting, whereas the ASRS immunity provisions remove some of these disincentives.

A similar design choice between cooperating with or separating from regulators is apparent in a comparison of ASAP with ASRS. American Airlines'

ASAP program pioneered an innovative way of overcoming barriers to reporting by building trust and cooperating with FAA regulators, rather than differentiating themselves from them, as in the ASRS model. The FAA does not grant immunity from enforcement action to ASAP program participants. However, if a pilot voluntarily reports a safety-related event that reveals an unintentional violation of an FAA regulation, the FAA responds with an administrative reprimand rather than punitive sanctions (Griffith, 1996).

Based on voluntary ASAP reports of inadvertent violations and other safety-related events, FAA regulators can learn about possible precursor conditions that otherwise might not have been reported, and the airline gets a detailed description of the conditions under which the event occurred. If similar events were reported to ASRS, the name of the airline would be de-identified; thus, the airline could not learn directly from the experience reported by its pilots. In ASAP, the FAA appears to have made a trade-off between using punitive means to enforce safety regulations and obtaining data necessary for the identification of precursors, and thus, improving safety conditions.

Health Care

Two health care reporting systems are modeled after ASRS: (1) the VHA Patient Safety Reporting System and (2) Applied Strategies for Improving Patient Safety (ASIPS). Although both systems are based on the ASRS model, they confront different legal barriers to reporting. Physicians employed by VHA hospitals are not subject to the same threat of litigation as physicians who practice in other settings. ASIPS, a Denver-based program, has modified the ASRS model to gather data on medical errors in ambulatory settings (e.g., doctors' offices). Unlike their colleagues in the VHA, members of the ASIPS collaborative are engaged in protecting their data from disclosure in litigation. The system designers, anticipating a legal challenge or security breach, are developing methods to ensure that serial numbers and computer identifiers cannot be used to link ASIPS reports to particular medical errors (Pace et al., 2003). Hence, the choice of confidentiality protections varies with the potential exposure to litigation or other threats.

POLICY IMPLICATIONS

As the examples from the aviation, nuclear power, and health care industries show, many types of organizations sponsor and support systems designed to identify precursors. These include government regulatory agencies (e.g., FAA and USNRC), individual organizations (e.g., airlines), hospital systems (e.g., VHA), industry associations (e.g., INPO), professional organizations (e.g., the Institute for Safe Medication Practices), and professional associations (e.g., Airline Pilots Association). Insurance companies could also contribute to precursor identification through activities designed to improve patient safety. For example, insurance companies could offer discounts in malpractice insurance to hospitals and physicians that participate in patient safety monitoring systems. They could also offer incentives to health care providers in hospitals and ambulatory settings to report close calls and identify precursors.

Three additional policy implications can be drawn from comparisons of industry efforts to identify accident precursors. First, based on the experiences of different industries, we can select better criteria for choosing design alternatives and evaluating and understanding the trade-offs involved in adopting alternative methods of detecting accident precursors. Second, we can gauge the strengths and weaknesses of systems and identify areas of expertise. For example, the aviation industry has developed several methods of detecting and gathering information about potential accident precursors. In addition, they have designed alternative models for aggregating data on different organizational levels, from the airline level to the level of air traffic control facilities to the interorganizational level encompassing everyone who uses the national airspace. Similarly, the nuclear power industry has demonstrated expertise in classifying and triaging events to identify accident precursors and weigh their probabilities.

Finally, we can conclude that every system design, whether organizational or interorganizational, requires trade-offs and has blind spots. No system can identify all of the conditions and behaviors that interact to produce disastrous events. To compensate for blind spots, we need multiple systems for precursor identification.

ACKNOWLEDGMENT

I wish to thank Marilyn Sue Bogner, Howard Kunreuther, and James Phimister for their insightful and instructive comments. Valuable support for this research was provided by the National Science Foundation, Decision Risk and Management Sciences Division (Grant SBR-9410749), and the Aetna Foundation's Quality Care Research Fund.

REFERENCES

Aviation Daily. 1996. FAA, American, Pilot Union to Begin 'Safety Program.' 316(15): 115.

Baker, H.M. 1997. Rules outside the rules for administration of medication: a study in New South Wales, Australia. Image: The Journal of Nursing Scholarship 29: 155–158

Bates, D.W., R.S. Evans, H. Murff, P.D. Stetson, L. Pizziferri, and G. Hripcsak. 2003. Detecting adverse events using information technology. Journal of American Medical Informatics Association 10: 115–128.

Battles, J.B., H.S. Kaplan, T.W. Van der Schaaf, and C.E. Shea. 1998. The attributes of medical event-reporting systems: experience with a prototype medical-event reporting system for transfusion medicine. Archives of Pathology and Laboratory Medicine 122: 231–238.

Brannigan, V., A. DiNovo, W. Freudenburg, H. Kaplan, L. Lakats, J. Minarick, M. Tamuz, and F. Witmer. 1998. The Collection and Use of Accident Precursor Data. Pp. 207–224 in Proceedings of Workshop on Accident Sequence Precursors and Probabilistic Risk Analysis, V.M. Bier, ed. College Park, Md.: University of Maryland Center for Reliability Engineering.

Callback. 2003. Cutoff on takeoff. Callback 286(July): 1. Available online at *http://asrs.arc.nasa.gov*.

Cohen, M., ed. 1999. Medication Errors. Washington, D.C.: American Pharmaceutical Association

Cullen, D.J., D.W. Bates, S.D. Small, J.B. Cooper, A.R. Nemeskal, and L.L. Leape. 1995. The incident reporting system does not detect adverse drug events: a problem for quality improvement. Joint Commission Journal of Quality Improvement 1:541-548.

Edlavitch, S.A. 1988. Adverse drug event reporting: improving the low U.S. reporting rates. Archives of Internal Medicine 148: 1499–1503.

Edmonson, A.C. 1996. Learning from mistakes is easier said than done: group and organizational influences on the detection and correction of human error. Journal of Applied Behavioral Science 32: 5–28.

Flight Safety Digest. 1998. Aviation safety: U.S. efforts to implement flight operational quality assurance programs. Flight Safety Digest 17(7-9): 1–56.

Griffith, S. 1996. American Airlines ASAP. Presentation at the Global Analysis and Information Network (GAIN) Workshop, October 22–24, 1996, Cambridge, Massachusetts.

Hardy, R. 1990. Callback: NASA's Aviation Safety Reporting System. Washington, D.C.: Smithsonian Institution Press.

Heget, J.R., J.P. Bagian, C.Z. Lee, and J.W. Gosbee. 2002. John M. Eisenberg Patient Safety Awards. System innovation: Veterans Health Administration National Center for Patient Safety. Joint Commission Journal of Quality Improvement 12: 660–665.

Heinrich, H.W. 1931. Industrial Accident Prevention. New York: McGraw-Hill.

Holtom, M. 1991. The basis for safety management. Focus on Commercial Aviation Safety 5: 25–28.

IOM (Institute of Medicine). 2000. To Err Is Human: Building a Safer Health System, L.T. Kohn, J.M. Corrigan, and M.S. Donaldson, eds. Washington, D.C.: National Academy Press.

La Porte, T.R. 1982. On the Design and Management of Nearly Error-Free Organizational Control Systems. Pp. 185–200 in Accident at Three Mile Island: The Human Dimensions, D.L. Sills, C.P. Wolf, and V.B. Shelanski, eds. Boulder, Colo.: Westview Press.

Leape, L.L. 1994. Error in medicine. JAMA 272: 1851–1857.

Levitt, B., and J.G. March. 1988. Organizational learning. Annual Review of Sociology 14: 319–340.

March, J.G., L.S. Sproull, and M. Tamuz. 1991. Learning from samples of one or fewer. Organization Science 2(1): 1–14.

Marcus, A.A., and M.L. Nichols. 1999. On the edge: heeding the warnings of unusual events. Organization Science 10(4): 482–499.

Minarick, J.W. 1990. The USNRC Accident Sequence Precursor Program: present methods and findings. Reliability Engineering and System Safety 27: 23–51.

National Academy of Public Administration. 1994. A Review of the Aviation Safety Reporting System. Washington, D.C.: National Academy of Public Administration.

NRC (National Research Council). 1980. Improving Aircraft Safety: FAA Certification of Commercial Passenger Aircraft. Washington, D.C.: National Academy of Sciences.

Pace W.D., E.W. Staton, G.S. Higgins, D.S Main, D.R. West, and D.M. Harris. 2003. Database design to ensure anonymous study of medical errors: a report from the ASIPS collaborative. Journal of American Medical Informatics Association 10(6): 531–540.

Perrow, C. 1984. Normal Accidents: Living with High-Risk Technologies. New York: Basic Books.

Reynard, W. 1994. Statement of Dr. William Reynard, director, Aviation Safety Reporting System, to the U.S. House of Representatives Subcommittee on Technology, Environment and Aviation, Committee on Science Space and Technology, July 28, 1994. Pp. 73–231 in 95-H701-21, testimony no. 2, Application of FAA Wake Vortex Research to Safety. Washington, D.C.: Congressional Information Service.

Reynard, W.D., C.E. Billings, E.S. Cheaney, and R. Hardy. 1986. The Development of the NASA Aviation Safety Reporting System, NASA Reference Publication 1114. Washington, D.C.: U.S. Government Printing Office.

Sagan, S.D. 1993. The Limits of Safety. Princeton, N.J.: Princeton University Press.

Stalhandske, E., J.P. Bagian, and J. Gosbee. 2002. Department of Veterans Affairs Patient Safety Program. American Journal of Infection Control 30(5): 296–302.

Tamuz, M. 1987. The impact of computer surveillance on air safety reporting. Columbia Journal of World Business 22(1): 69–77.

Tamuz, M. 2000. Defining Away Dangers: A Study in the Influences of Managerial Cognition on Information Systems. Pp. 157–183 in Organizational Cognition: Computation and Interpretation, T.K. Lant and Z. Shapira, eds. Mahwah, N.J.: Lawrence Erlbaum Associates.

Tamuz, M. 2001. Learning disabilities for regulators: the perils of organizational learning in the air transportation industry. Administration and Society 33(3): 276–302.

Tamuz, M., H.S. Kaplan, and M.P. Linn. 2001. Illuminating the Blind Spots: Studying Organizational Learning about Adverse Events in Blood Banks. Presented at the Academy of Management Annual Meeting, August 2001, Washington, D.C.

Tamuz, M., E.J. Thomas, and K.E. Franchois. 2004. Defining and classifying medical error: lessons for patient safety reporting systems. Quality and Safety in Health Care 13: 3–20.

Tasca, L. 1989. The Social Construction of Human Error. Ph.D. Dissertation. State University of New York-Stony Brook.

Thomas, E.J., and R.L. Helmreich. 2002. Will Airline Safety Models Work in Medicine? Pp. 217–232 in Medical Error, K.M. Sutcliffe and M.M. Rosenthal, eds. San Francisco: Jossey-Bass.

Vincent C., N. Stanhope, and M. Crowley-Murphy. 1999. Reasons for not reporting adverse incidents: an empirical study. Journal of Evaluation in Clinical Practice 5: 13–21.

Weick, K.E. 1995. Sensemaking in Organizations. Thousand Oaks, Calif.: Sage Publications.

Defining and Analyzing Precursors

WILLIAM R. CORCORAN
Nuclear Safety Review Concepts Corporation

History is a vast early warning system.
—*Norman Cousins*

Wouldn't it be nice if we could identify precursors before they "precursed" disasters? With great retrovisual acuity, experts and laypeople alike can identify the precursors to *Challenger*, *Concorde*, Three Mile Island (TMI), Davis-Besse nuclear power plant in 2002, *Columbia*, and other consequential adverse events.

Consider the space shuttle *Challenger*. We now know that every shuttle launch that included an O-ring blow-by before the *Challenger* explosion was a precursor to an explosion in that if the pre-launch ambient temperature had been sufficiently low the O-rings would have failed and the vehicle would have been lost.

In the case of the supersonic airplane *Concorde*, an examination of the accident history indicates about a half-dozen recorded precursors to the fatal encounter with a foreign object. These precursors involved takeoffs with either foreign objects on the runway or tire blowouts or both. And what about the unrecorded precursors? For instance, were there unrecorded times when *Concorde* took off when there was a foreign object on the runway? Might these events have been precursors, even though we don't know about them?

In the case of TMI, we now know that every case of a stuck-open, power-operated relief valve (PORV) that occurred before the accident was a precursor to a potential core meltdown. However, before TMI, few, if any, nuclear reactor engineers would have believed that operators would fail to recognize the symptoms of a stuck-open relief valve; nor would they have believed that operators would reduce makeup flow in the face of symptoms of inadequate coolant inventory.

Has there ever been a serious, consequential adverse event that did not have precursors? Chernobyl and the *Hindenburg* were said to have come "out of the

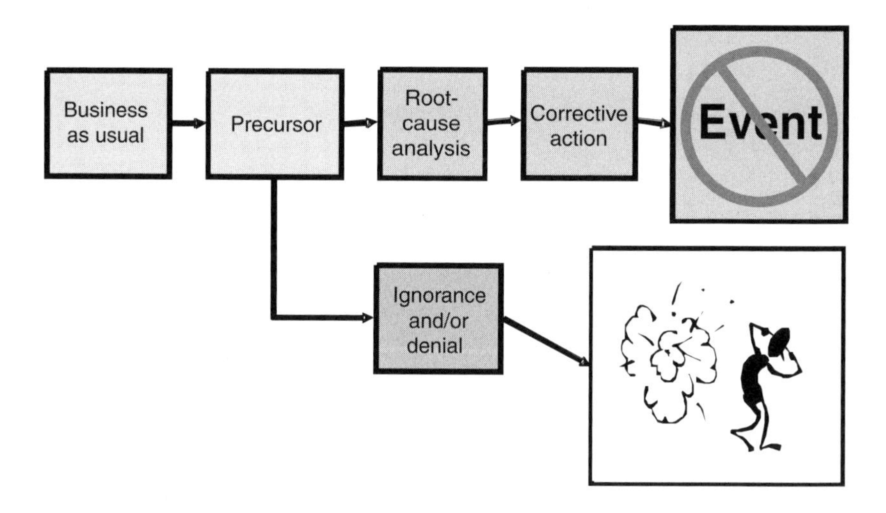

FIGURE 1 Root cause analysis (RCA) and corrective action (CA) after identification of a precursor can prevent a consequential event.

blue," but did they? Would sufficient access to the history of these events reveal precursors that, had they been recognized and attended to, might have averted them? An old cowhand might ask, "Why not head them off at the pass?" That is to say, why not identify and analyze the precursors and take corrective action to prevent the downstream consequential adverse event (Figure 1).

WHAT ARE PRECURSORS?

The National Academy of Engineering workshop definition of an accident precursor is any event or group of events that must occur for an accident to occur in a given scenario. One dictionary definition (among many) is "one that precedes and indicates the approach of another." For the purpose of this paper, a precursor is defined as a situation that has some, but not all, of the ingredients of a more undesirable situation. Thus, a precursor is an event or situation that, if a small set of behaviors or conditions had been slightly different, would have led to a consequential adverse event. Has there ever been a consequential event, near miss, or infraction/deviation that did not have a precursor? In some sense of the word, probably not. Have there been consequential events with precursors that have been discounted, dismissed, not recognized, or not understood? Most certainly.

What Keeps a Precursor from Being a Real McCoy?

The "real McCoy" in this case is, of course, a highly consequential adverse event. When less than a real McCoy happens, the real McCoy does not occur for one of three reasons: (1) an exacerbating factor was missing; (2) a mitigating factor was effective; or (3) both.

To express these three ideas as equations, we have:

$$\{\text{Real McCoy}\} = \{\text{Precursor}\} + \{\text{Exacerbating Factor(s)}\} \quad (1)$$

Equation 1 says that, if the next occurrence of the precursor includes specific exacerbating factors, a consequential event will result.

$$\{\text{Real McCoy}\} = \{\text{Precursor}\} - \{\text{Mitigating Factor(s)}\} \quad (2)$$

Equation 2 says that, if the next occurrence of the precursor situation does not include important defenses, barriers, or other mitigating measures, a consequential event will result.

$$\{\text{Real McCoy}\} = \{\text{Precursor}\} + \{\text{Exacerbating Factor(s)}\} - \{\text{Mitigating Factor(s)}\} \quad (3)$$

Equation 3 combines the thoughts of Equations 1 and 2.

Can Real McCoys Be Precursors?

As was recently illustrated, a real McCoy can be a precursor, too. On January 8, 2002, at St. Raphael Hospital in Connecticut, a woman was killed in an operating room when she was given nitrous oxide instead of oxygen. Three days later, another woman was killed in the same operating room in the same way, thus providing a tragic example of not learning from experience. Precursors of this type can be expressed by Equation 4:

$$\{\text{Real McCoy}\}_{(N+1)} = \{\text{Real McCoy}\}_{(N)} + \{\text{Nothing}\} + \{\text{Time}\} \quad (4)$$

This equation says that, if an adverse event is not effectively investigated and appropriate corrective action taken, the causes of the event may continue to exist. And as long as the causes continue to exist, a similar event may occur. Examples of this type include the infamous Ford Explorer-Bridgestone/Firestone episode and the tragedies of Therac-25, a radiation therapy accelerator.

Real McCoys might also be considered precursors using Equation 4a:

$$\{\text{Worse Real McCoy}\}_{(N+1)} = \{\text{Real McCoy}\}_{(N)} + \{\text{Nothing}\} + \{\text{Time}\} + \{\text{Exacerbating Factor(s)}\} \quad (4a)$$

An example of this was the loss of some of the crew of the USS *Squalus* (SS-192), which was a precursor to the loss of the entire crew of the USS *Thresher* (SSN-593). Both submarines sank because of loss of hull integrity. The real McCoy and the precursor are related by both Equation 4 and Equation 2, which together are captured in Equation 4a. If real McCoys are also precursors that indicate the approach of a downstream real McCoy, wouldn't prudent people take action to head them off at the pass?

Near Misses

A near miss is a special kind of precursor (some people like to say "near hit" or "close call" for the same concept). In general, we think of a near miss as a precursor with ingredients that differ in only minor or non-robust ways from those necessary for a consequential event. For instance, when the necessary exacerbating factors are highly likely, the precursor is called a near miss. For example, running a red light in a busy intersection without causing a collision is a near miss. The exacerbating factor would have been another vehicle crossing the intersection. Similarly, one would expect a precursor to be called a near miss if the mitigating factors were unlikely or not robust. For example, a steam pipe break that does not result in injuries because the workers happen to be at lunch when it happens could be considered a near miss. (This actually happened at Millstone Unit 2 in the mid-1990s.) The near miss concept suggests the following:

$$\{\text{Real McCoy}\} = \{\text{Near Miss}\} +/- \{\text{Not Much}\} \quad (5)$$

Many people believe that investigations of near misses should be commensurate with investigations of the corresponding averted consequential events. Thus, many shuttle launches prior to *Challenger* and *Columbia* were "secret" near misses. Some *Concorde* accidents before the fatal one were also "secret" near misses.

Managers and program people should be asking what kept a near miss from being worse and how close it came to being a real McCoy. Perhaps, in the cases of *Challenger* and *Concorde*, the near misses were not obvious or fully appreciated as precursors.

Unveiling Precursors

If it were known that a specific event was a precursor of an accident, people would certainly do something to avert the next real McCoy. This is almost a tautology, but it needs to be said. However, many precursors that should indicate the approach of a real McCoy are not recognized. For example, *Concorde* program personnel kept records of precursors involving *Concorde* aircraft, but apparently they did not "connect the dots" to envision an encounter with a foreign object on takeoff that could destroy an aircraft. Precursors to *Challenger* (O-ring blow-by) and *Columbia* (foam strikes) also went unrecognized.

Notice that all of the "postcursor" real McCoys mentioned in this paper were preceded by precursors that did not sufficiently indicate their approach. If the precursors had been "unveiled" for the threats they indicated, the accidents might have been averted. To unveil something is to reveal its true nature, and clearly lives, pain, assets, and careers could be saved if organizations could unveil precursors. *People* unveil precursors when they make inferences from events and situations (because events and situations are not capable of implying anything on their own). One systematic approach to making inferences from potential precursor events and situations is root-cause analysis, which can be helpful in deconstructing events and situations to aid decision making.[1]

ROOT-CAUSE ANALYSIS

In applying root-cause analysis to possible precursor events and conditions, two questions must be considered: (1) how does one select events and situations as potential precursors; and (2) how does one perform a root-cause analysis on selected events and situations. Before a precursor can be analyzed, it must be recognized as an ingredient in a recipe for dire consequences. If today's anomaly or today's usual practice cannot be envisioned as an ingredient in such a recipe, there is no hope that it will be unveiled or detected.

For example, at Davis-Besse, a U.S. nuclear power plant, there were dozens of anomalies that were recognized, in retrospect, as ingredients in a recipe for an extended shutdown. When one anomaly is not recognized as a precursor, the failure can be explained as a narrow gap in knowledge. But when dozens of anomalies are not recognized, one begins to wonder about programs, processes, organizations, interfaces, and, of course, safety culture. At Davis-Besse, as reported in local newspapers, there were many precursors:

- leaky control-rod drive-mechanism joints that encouraged tolerance of leakage
- boric acid deposits in the reactor vessel head area from leaks
- the presence of alloy 600, which is subject to cracking
- time, temperature, and stress
- criticism by the Nuclear Regulatory Commission (that was ignored) of the boric acid corrosion-control program
- predictions by an industry group that cracks were likely

[1]For Internet access to a large community of root cause analysis practitioners, as well as links, files, database tables, and other resources, see *http://groups.yahoo.com/group/Root_Cause_ State_of_the_Practice/*.

- small cracks (not knowable directly)
- boric acid issuing from the small cracks (not knowable directly)
- difficulties in inspecting reactor vessel heads because of the design
- truncated inspections
- disapproval of proposed changes to facilitate inspections
- news of alloy 600 cracks in similar plants
- increase in rust-colored boric acid deposits
- clogging of radiation-monitor system filters
- fouling of containment air-cooler heat-transfer surfaces
- failure to do root-cause analyses of any of these anomalies
- falsified auditing of the boric acid corrosion-control program
- poor regulatory and industry oversight

How can events and conditions be understood as possible precursors? Equations 1–5 can be a helpful starting point. For serious real McCoy situations that match Equations 4 and 4a, such as *Squalus* and the first St. Raphael fatality, the root-cause analyses should include consideration of these events as precursors. There are obvious lessons to be learned, and these events should have been examined for (1) their potential for being repeated and (2) their potential of being repeated *and* being worse.

In the case of the submarine *Squalus*, half of whose crew was rescued after the vessel sank, the investigation did not focus sufficiently on the factors that had kept the consequences from being worse. Understandably so. The *Squalus* accident was the first time the submarine rescue system (the basis for today's submarine rescue systems) was used. The investigation, however, did not result in advising submarine commanders to choose test sites sufficiently deep to achieve test objectives but shallow enough to avoid collapsing unflooded compartments.

This is a special case of an important safety principle that tells us not to take risks in excess of those for which there is some benefit (see Corcoran [2002] for a list of the safety principles). Accidents that are narrowly averted (near misses as described by Equation 5) should be examined as precursors, focusing on the factors that kept the consequences from being worse.

It is more difficult to recognize as precursors events and situations with less obvious similarities to accidents. To assist in recognizing these, the event or situation should be considered in terms of Equations 1–3, which indicate the potential likelihood and severity of the possible accident to determine if the event should be considered for further precursor analysis.

Root-Cause Analysis of Precursors

Anecdotal experience suggests that the difference in occurrence rates between the levels of severity of accidents, near misses, compromises, and infractions is

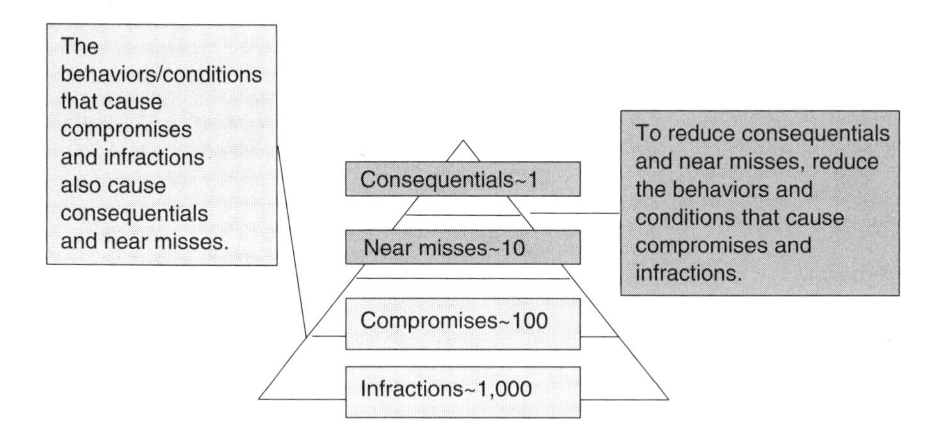

FIGURE 2 The occurrence pyramid.

about a factor of 10 (Figure 2). Experience with root-cause analysis indicates that, as a general rule, the causes of compromises, infractions, and deviations are the same as the causes of near misses and consequential events. And, investigations of consequential events and near misses show that higher severity events include an accumulation of lower level events and causes. Hence, root-cause analyses of the precursors to accidents should help reduce accident rates.

Root-cause analysis is commonly performed on consequential events, although it can also be performed on low-consequence precursors. In all cases, the analysis is based on evidence and goes deep enough to reveal important underlying issues, while ensuring that chains of influence are tightly linked and pursuing the generic implications of causes and effects. In performing a root-cause analysis, eight questions can be applied to accidents, incidents, and near misses (Corcoran, 2002). The first two questions consider the outcome that events might be repeated or might occur as accidents. Questions 3–6 consider influences on outcomes and the factors that limited, controlled, or restricted the consequences. Questions 7 and 8 are meant to "close out" an analysis and risk reduction, which cannot be achieved without implementing corrective actions.

Question 1. What were, are, or will be the consequences of the potential precursor? Consequences are adverse outcomes of events. As defined by Equation 4, if nothing is done following an event, it may become a precursor to a similar event in the future with similar consequences. Consequences that should be examined in a root-cause analysis are the actual consequences that have accrued to date, the expected consequences in the pipeline, and potential consequences that have so far been averted by the absence of exacerbating factors and/ or the presence of mitigating factors.

Question 2. What does the event mean or signify to the victims and other stakeholders? The significance of an event includes potential consequences (mentioned in Question 1) and how the occurrence of that event would impact the stakeholders.

Question 3. What vulnerabilities set the stage for the consequences? If the situation had not been set up for the event, it could not have happened. To analyze precursors, you must define "the recipe" for the consequences to occur.

Question 4. What triggers or initiates a chain of events? Vulnerability alone does not cause consequences. It takes a trigger or initiating action. For example, what were the triggers for *Concorde*? Some might say the trigger was a previous aircraft dropping a foreign object on the runway. Others might say it was the takeoff roll-out itself that triggered the accident. Some triggers can be considered precursors in and of themselves.

Question 5. What makes the consequences as bad as they are? In some cases, vulnerability and the trigger alone do not cause the consequences of interest. Something else exacerbates the situation, amplifying the adverse effects or continuing the damaging mechanism or the like.

Question 6. What kept or is keeping the consequences from being worse? In the vast majority of consequential events, and in all near misses, there were factors that limited, controlled, or restricted the consequences. For example, the 2002 Davis-Besse situation did not become a loss-of-coolant accident because degradation of the reactor vessel head was discovered during repair of a crack in the nozzle.

Question 7. What should be learned from the event? Answering this question determines the lessons to be learned, the factual basis of each lesson to be learned, and who should learn the lesson.

Question 8. What should be done about it? To avert the consequences of the future real McCoys indicated, suggested, or announced by precursors, corrective actions must include not only controlling the precursor behaviors and conditions, but also controlling the processes that produce them. In determining corrective actions, the chains of causation must be interrupted. The causal events relate to (1) what set up the situation, (2) what triggered the event, and (3) what made the event as bad as it was.

Tools

Several tools are available to assist in answering these questions (Corcoran, 2003a,b). For instance, to help with Questions 3–6, a Comparative TimeLine© can be used to organize data. Graphically oriented approaches, such as event and causal-factors charts, can be useful for laying out events. Staircase trees can be used to establish chains of influence.

For answering Questions 7 and 8, tables and matrices can help make sense of influences. Some useful tools include: the missed-opportunity matrix, the barrier-analysis matrix, the cause-consequence matrix, the lessons-to-be-learned matrix, and the regulatory-infraction matrix.

CONCLUSION

Severe adverse events "from out of the blue" (i.e., accidents without precursors) are rare. Detailed investigations of most adverse events reveal precursors—that is, accidents have been preceded by events, behaviors, and conditions that were ingredients of the recipe for the adverse consequences. Adverse events that seem to come out of the blue are events whose precursors were not recognized.

The ability to recognize precursors and respond appropriately is a very valuable organizational skill—especially the ability to identify or unveil precursors. Unless the precursor nature of an event, behavior, or condition is recognized, it is not likely to get much attention. Almost as important as unveiling precursors is recognizing the generic implications of events (if this happened [or existed], what else could one expect?).

Organizations must prioritize precursors. Addressing precursors that are departures from regulatory requirements must be a high priority. Precursors that constitute immediate threats to life or health must also be attended to promptly. Precursors that may be ingredients of complex accident recipes whose outcomes are not fully understood are harder to prioritize. Suffice it to say that prioritizing precursors is not a trivial task.

Clearly, it would be helpful if adverse events were reported transparently so the fragility of a situation implied by cumulative precursors could be understood. TMI, *Challenger*, *Concorde*, *Columbia*, Davis-Besse 2002, the Millstone regulatory shutdown, and other events may well have been averted if the fragility of the situations that led to them had been known to accountable individuals.

REFERENCES

Corcoran, W.R., ed. 2002. Firebird Forum 5(7). Available online at *http://groups.yahoo.com/group/Root_Cause_State_of_the_Practice/*.

Corcoran, W.R. 2003a. The Phoenix Handbook. Windsor, Conn.: Nuclear Safety Review Concepts Corporation.

Corcoran, W.R. 2003b. Firebird Forum 6(1). Available online at *http://groups.yahoo.com/group/Root_Cause_State_of_the_Practice/*.

Nuclear Accident Precursor Assessment

The Accident Sequence Precursor Program

MARTIN B. SATTISON
Idaho National Engineering and Environmental Laboratory

The U.S. Nuclear Regulatory Commission (USNRC) has operated an accident precursor program since 1979, pioneering this particular field of safety. The Accident Sequence Precursor (ASP) Program has provided useful insights into the effects of operational events on safety in the nuclear industry ever since. In the past 24 years, the program has matured along with the risk assessment tools and models upon which it depends.

The first probabilistic risk assessment (PRA) by a commercial nuclear power plant, the Reactor Safety Study (WASH-1400), was completed in 1975 (USNRC, 1975). The USNRC formed the Risk Assessment Review Group (commonly referred to as the Lewis Committee) to perform an independent evaluation of WASH-1400. That committee made a number of recommendations in 1978, including that more use be made of operational data to assess the risk from nuclear power plants. The Review Group's report stated, "It is important, in our view, that potentially significant (accident) sequences, and precursors, as they occur, be subjected to the kind of analysis contained in WASH-1400" (USNRC, 1978). In response to that recommendation, the USNRC's Division of Risk Analysis established the ASP in the summer of 1979, shortly after the Three Mile Island (TMI-2) accident. The first major report of that program, *Precursors to Potential Severe Core Damage Accidents: 1969–1979, A Status Report (NUREG/CR-2497, Volume 1)*, was formally released in June 1982 (Minarick and Kukielka, 1982).

The primary focus of ASP was on evaluating the risk for a specific time period from all operating nuclear power plants (not individual plants), and this is still a primary objective of ASP. The implications of this objective have significantly influenced the way analyses are done, the nature of the results, and the types of insights expected from the program.

In the early years of ASP little emphasis was placed on detailed, accurate, plant-specific risk models. To the contrary, early risk models were generic and did not differentiate the physical and operational characteristics of plants in the nuclear fleet, except on a very crude level. However, this level of detail was adequate for the purposes of determining and trending the risk of high-level, industry-wide, severe core damage. In fact, ASP was pushing the state of the art in risk assessment to the limit, and asking for more would have been impractical. Only a few risk models besides WASH-1400, such as those for the Zion and Indian Point nuclear power plants, could have supported detailed risk assessments of operational events.

The first ASP risk models consisted of two sets of standardized, functional event trees, one for pressurized water reactors (PWRs) and one for boiling water reactors (BWRs), the two unique reactor designs in the U.S. light water reactor fleet. Each set of event trees presented the accident sequences stemming from four initiating events selected for the study:

- loss of main feed water (the system that extracts the heat from the reactor)
- loss of off-site power (requiring alternative power sources for key safety equipment)
- small loss-of-coolant accident (LOCA) (a direct leak of coolant from the reactor pressure boundary)
- break in the steam line (requiring actions to control reactivity and establish alternative long-term heat removal)

The first two initiating events were considered the most likely off-normal events of concern; the latter two represented bounding events for many of the safety-related systems in a reactor plant. The event trees were used to model most of the events selected as precursors. Figure 1 shows the standard event tree for loss of main feed water in a PWR (Minarick and Kukielka, 1982). With this limited set of event trees, a number of events of interest could not be properly evaluated without additional work. In these cases, unique event trees were developed.

Accident precursors were quantified in the framework of the event trees. "Unusual" initiating events and complete failures of safety-related functions were selected as precursors. The frequencies of initiating events were calculated based on the operating experience of the plants from 1969 through 1979. Function failure (branch point) probabilities were calculated based on observed failures in the operating event data and estimates of the number of test demands and additional nontest demands to which the function would be expected to respond. For each precursor event, the appropriate values were applied to the event tree accident sequences for which the observed event was considered a precursor. Because operators would not just sit back and watch an accident progress, the

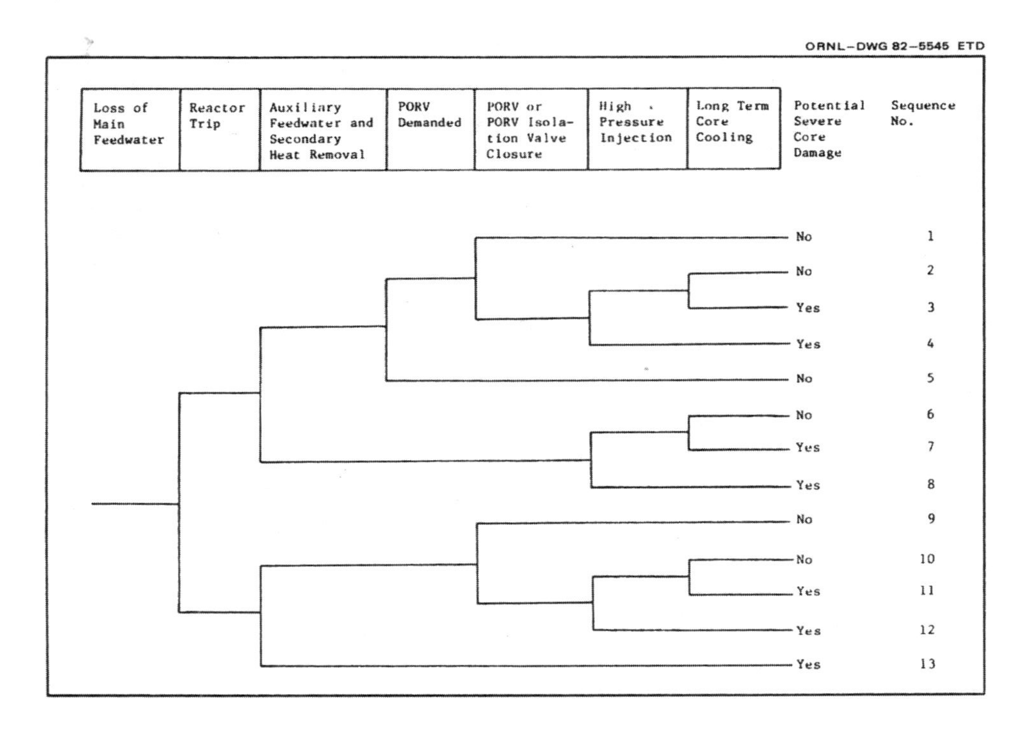

FIGURE 1 Standard event tree for loss of main feed water in a PWR. Source: Minarick et al., 1982.

chance that a failure or initiating event could be rectified was included as a recovery action. The same process was used to evaluate the 1980–1981 accident precursors (Cottrell et al., 1984).

The event tree models used for assessments from 1969 to1981 were acknowledged to be less than adequate. In 1984, ASP, with the help of the USNRC Accident Sequence Evaluation Program, identified classes of plants based on common responses to specific initiating events (transient, loss of off-site power, and small LOCA) and began to develop computerized, systemic event trees for each plant class. Based on the structure of the event trees, four PWR and three BWR plant classes were created. System models based on the train-level configurations were used in conjunction with plant-class event sequences to distinguish differences among plant designs.

The PWR reactor-trip event trees from the 1985 report (Minarick et al., 1986) were better representations of the individual plants, but still allowed use of operational data at a higher level (plant class). If a model is too plant specific, there are not enough operational data to evaluate events with confidence. If the model is too generic, the insights from the data will be limited. The 1985 models reflected a shift in emphasis away from industry averages toward the identification of specific precursors and a determination of their significance.

The new models were first used to evaluate (in parallel) the 1984 and 1985 accident precursors. In 1984, to reduce the time between the occurrence of precursors and their analysis, ASP skipped the 1982 and 1983 events and began analysis of the 1984 and 1985 events. The 1985 ASP report came out in December 1986 (Minarick et al., 1986); the 1984 ASP report came out in May 1987 (Minarick et al., 1987). The 1986 precursors were evaluated using essentially the same models.

Revised models were used to evaluate the 1987 precursors. The definitions of the BWR plant classes were adjusted; and the La Crosse plant was different enough to be placed in its own class. The PWR plants were divided into eight classes instead of four, which made possible better representations of actual plant configurations and characteristics (Minarick et al., 1989). Other changes were made to reflect new data on operator performance, to enable better models of emergency battery depletion during a station blackout (a total loss of all AC power sources), and to require that operable water-injection sources be available during venting of the containment building. (All commercial reactors in the United States are surrounded by containment buildings, which add another barrier between reactors and the environment.)

Models for the 1988 precursors were again significantly changed. The plants were grouped into eight classes: three for BWRs and five for PWRs. Core vulnerable sequences from previous models were reassigned, either as success or as core damage, and the likelihood of a failure of a reactor coolant pump seal following a station blackout was explicitly modeled. These models were used for precursor evaluations from 1989 through 1993.

The 1992 and 1993 precursor analyses included the potential use of alternate equipment and procedures, beyond those considered in the basic risk models, which had recently been added by licensees to provide additional protection against core damage. The 1992 precursor analyses were the first event assessments reviewed by plant licensees before they were published. This process has continued ever since. The 1982 and 1983 ASP evaluations were not begun until 1994 and were not completed until 1997. The same models and methods were used in 1993.

The 1994 ASP evaluations were the first to use models developed especially for ASP (Sattison et al., 1994) using the SAPHIRE risk assessment software package (USNRC, 1995). The events trees were expanded to include significantly more detail and additional initiating events, such as rupture of a steam-generator tube and anticipated transients without SCRAM (reactor trips). Plant-specific fault trees were used to capture the unique features of plant systems. Seventy-five plant-specific models were used to analyze precursors for the entire commercial fleet of more than 100 reactors.

The SAPHIRE-based models, which have been used ever since the 1994 analyses, have been improved based on visits with risk staffs at each facility. Changes were also made, and are still being made, in response to peer reviews.

THE EVENT SCREENING PROCESS

The nuclear industry was uniquely positioned to start an accident-precursor program because an operational data-collection mechanism, mandated by law, was already in place. In accordance with the U.S Code of Federal Regulations (10CFR50), commercial nuclear power plants are required to report to the USNRC all operational events that represent a deviation from the licensing basis or failure/degradation of a safety function. The USNRC has permanent, on-site resident inspectors at each nuclear power plant to oversee daily activities. Thus, failure to report as required by law can be readily detected. In addition, potential penalties are severe, so compliance is virtually absolute. Reports submitted to NRC to satisfy the law, called licensee event reports (LERs), have a standard format and very detailed guidelines. LERs are closely scrutinized by the USNRC, and anything unclear is questioned and resolved with the submitter. LERs are then screened for a number of programs, including ASP. The USNRC determined that the reporting criteria established in the CFR would ensure that ASP could capture most potential accident precursors.

In the first ASP report, about 19,400 LERs were examined for accident precursors, which were defined in general terms as "events that are important elements in a chain of events (an accident sequence) possibly leading to core damage. Such precursors might be infrequent initiating events or equipment failures that, when coupled with one or more postulated events, could result in a plant condition leading to severe core damage" (Minarick and Kukielka, 1982).

There were specific acceptance criteria for further evaluation of accident precursors:

- any failure of a system that should have functioned as a consequence of an off-normal event or accident
- any instance of two or more failures
- all events that resulted in or required initiation of safety-related equipment (except events that required only a reactor trip and the reactor trip was successful)
- all complete losses of off-site power and any less-frequent, off-normal initiating events or accidents
- any event or operating condition that was not within the plant design bases or that proceeded differently from the plant design bases
- any other event that, based on the reviewer's experience, could have resulted in or significantly affected a chain of events leading to potential severe core damage

These criteria served only as guidelines, and the reviewers were heavily relied upon to exercise judgment during the screening process to ensure that no LERs were screened out that shouldn't have been and that the screening process effectively reduced the number of events requiring further evaluation. The evaluation of the 1980–1981 events used the same acceptance criteria and processes.

The first changes to the LER selection process were made for the selection of 1985 precursors (the next events analyzed after the 1980–1981 events). The six criteria listed above were consolidated and simplified down to five criteria with little change in their meaning. The selection process was separated into two parts, an initial screening and a detailed review. In addition, more consideration was given to events that could not be easily categorized:

- flooding and fire
- water hammer (a thermal-hydraulic phenomenon that can cause high stresses in piping)
- natural phenomena, such as earthquakes and tornadoes
- inadvertent activation of safety systems
- natural circulation degradation (coolant circulation caused by differences in temperature and elevation)
- failures of control systems
- reactivity insertion (changes in the ability to sustain a controlled nuclear chain reaction)
- inadvertent closure of the main steam-isolation valve (interference with the normal method of removing heat from the reactor)
- excessive coolant or steam generator inventory

One of the major changes to the precursor selection process was a revision to the LER rule that became effective in 1984 requiring that a detailed report be provided of all operational events involving a reactor trip. All of these events were captured in the initial screening.

The 1987 precursor selection process, in addition to the typical precursors, identified events involving a loss of containment function and other events that were considered serious but were not modeled (although these were not called precursors). Two changes were made to the LER review and precursor selection process. First, LERs were initially prioritized for further review using the Sequence Coding and Search System (SCSS) database to identify the candidate LERs. Second, events were included in the main body of the report only if they had conditional core-damage probabilities (CCDPs) greater than 1×10^{-6} per reactor year. This was the first time ASP used quantitative criteria in the precursor determination process.

The screening and review process for the 1988–1992 precursors was significantly modified. The initial screening of LERs was performed by the USNRC Office for Analysis and Evaluation of Operational Data (AEOD), which used criteria that were more oriented toward regulatory and safety issues and less oriented toward risk.

The 1993 analysis used the screening criteria from the previous year, but added the criterion that any event must be included for which an augmented inspection team (AIT) or incident investigation team (IIT) report was written. AITs and IITs are formed for events of special interest or significance to USNRC. The same criteria were used for the 1994–1997 analyses, as well as for selection of the 1982–1983 precursors, which were not completely analyzed until 1997.

Little has changed in the selection criteria since the 1993 precursor report. In the late 1990s, it was recognized that long-duration unavailability of a key component, even if there was sufficient redundancy, could be risk-significant. The screening algorithm for the SCSS review was therefore revised to capture these types of events.

TRENDS OF THE RESULTS

For the 1969–1979 analyses, 169 accident sequence precursors were identified. The frequency of severe core damage was estimated to be 2.3×10^{-3} per reactor year. The results did not show any variation with plant age, plant type, plant capacity, vendors, or architect-engineers.

The analysis of 1980–1981 operational events represents a transitional period immediately following the TMI-2 accident. During that period, many plant configuration and operational changes were mandated, with implementation taking place over a long transitional period. Fifty-eight events were selected as accident precursors, approximately the same number per year as in 1969–1979, but the risk significance of these events was less. Lower risk was attributed to improvements in some system reliabilities, additional protective features, and a decrease in the degree of coupling observed in the precursors. The estimated industry average frequency of severe core damage based on the 1980–1981 precursors was 1.6×10^{-4} per reactor year. (The 1982 and 1983 precursors were not analyzed until 1997 and will be discussed later.)

The 1984 and 1985 precursors were analyzed in parallel; the 1985 precursor report came out six months before the 1984 report. The 1986 precursors used the same models and methods (Minarick et al., 1988). Forty-eight precursors were identified in 1984, 63 in 1985, and 34 in 1986. The 1984, 1985, and 1986 reports did not present an average severe core-damage frequency based on the precursors. Instead, distributions of precursors as a function of CCDP were shown in a table (Table 1).

In the evaluation of the 1987 precursors, ASP began to distinguish between precursors with a CCDP greater than 1×10^{-6} and those with a CCDP of less than 1×10^{-6}; only the former were included in the main body of the report. In 1987, there were 63 precursors, 33 of which had a CCDP of 1×10^{-6} or higher. The 1988 precursor report identified 32 precursors (greater than 1×10^{-6}) (Minarick et al., 1990a). The 1988 report also identified 28 LERs that were impractical to analyze but were described in a table. The 1989 precursor report identified 30 precursors and 27 events that were potentially significant but impractical to analyze (Minarick et al., 1990b).

The 1990 precursor report identified 28 precursors and 53 events that were potentially significant but impractical or lacked sufficient information to analyze (Minarick et al., 1991). In 1990, for the first time, two events were analyzed that took place while the reactor was shut down. The 1991 precursor report (Minarick et al., 1992) identified 29 precursors and 45 events that were potentially signifi-

TABLE 1 Distribution of Precursors as a Function of CCDP, 1984–1986

	Number of Precursors		
CCDP	1984	1985	1986
1×10^{-2} to 1×10^{-1}	0	1	0
1×10^{-3} to 1×10^{-2}	1	1	2
1×10^{-4} to 1×10^{-3}	16	8	4
1×10^{-5} to 1×10^{-4}	8	14	8
1×10^{-6} to 1×10^{-5}	8	16	5
1×10^{-7} to 1×10^{-6}	8	7	3
1×10^{-8} to 1×10^{-7}	3	7	7
1×10^{-9} to 1×10^{-8}	2	6	4
1×10^{-10} to 1×10^{-9}	2	3	1

TABLE 2 Distribution of Precursors as a Function of CCDP, 1987–1997

	Number of Precursors										
CCDP	1987	1988	1989	1990	1991	1992	1993	1994	1995	1996	1997
1×10^{-2} to 1×10^{-1}	0	0	0	0	0	0	0	0	0	0	0
1×10^{-3} to 1×10^{-2}	0	0	0	0	1	0	0	1	0	1	0
1×10^{-4} to 1×10^{-3}	10	7	7	6	13	7	4	1	1	2	0
1×10^{-5} to 1×10^{-4}	9	14	11	11	8	7	7	4	7	4	2
1×10^{-6} to 1×10^{-5}	14	11	12	11	6	13	5	3	2	7	3
$< 1 \times 10^{-6}$	30	—	—	—	—	—	—	—	—	—	—

cant but impractical to analyze. The 1992 precursor report identified 27 precursors and 51 events that were potentially significant but were considered impractical to analyze (Copinger and Mays, 1993). The 1993 precursor report identified 16 precursors and 19 events that were potentially significant but impractical or lacked sufficient information to analyze (Vanden Heuvel et al., 1994).

The 1994 precursor report identified eight at-power precursors, one shutdown precursor, one containment-related event, nine "interesting" events, and twelve potentially significant events considered impractical to analyze (Belles et al., 1995). Interesting events are not usually precursor events although they shed light on unusual failure modes with the potential to compromise core cooling.

The 1995 precursor report identified ten at-power precursors, no shutdown precursors, no containment-related events, six interesting events, and one potentially significant event considered impractical to analyze (Belles et al., 1997a). The 1996 precursor report identified 13 at-power precursors, one shutdown precursor, no containment-related precursors, two interesting events, and two potentially significant events considered impractical to analyze (Belles et al., 1997b). The 1997 precursor report identified five at-power precursors, no shutdown precursors, no containment-related precursors, five interesting events, and no potentially significant events considered impractical to analyze (Belles et al., 1998). The distributions for 1987–1997 are shown in Table 2.

Because of the wide-ranging changes in the event-selection criteria and processes, changes in modeling methods, and the increasing sophistication of the risk models, it would not be appropriate to determine overall risk trends from ASP reports.

LESSONS LEARNED

ASP was established fairly early in the development of probabilistic risk assessment for the commercial nuclear power industry. The driving force behind the program was the potential benefits of risk assessment and learning from operating experience (based on the first comprehensive, probabilistic risk assessment, WASH-1400).

ASP was established just at the time of the TMI-2 accident, which was unfortunate for two reasons. First, because of TMI-2 there was a sense of urgency about getting the program started to see if other potential TMIs were lurking about. This emphasis caused the program to become focused on a narrow range of issues rather than exploring broader goals, such as the classification and ranking of precursors according to frequency of occurrence rather than CCDP. Second, TMI-2 forced ASP in the direction of post-event risk assessment. The primary questions after TMI-2 were the probability of another TMI-2, whether existing probabilistic risk assessments could have predicted TMI-2, and how close other events had come to causing core damage.

Ideally, a comprehensive accident precursor program should accomplish a number of goals:

1. Identify the nature of accident precursors for the industry. This requires that precursor categories be defined based on accident sequences determined from full-scope risk assessments for the entire range of facilities and systems. This is important because accident precursors are typically small segments of one or more accident sequences, and assessing accident precursors includes mapping these events onto the risk models. If noteworthy events are observed that cannot be mapped, the risk models may not be adequate.
2. Prioritize or rank precursor categories based on both frequency of occurrence and risk significance. Ranking by frequency of occurrence for each category of precursor indicates the weaknesses in facilities at risk for accidents. Ranking by risk significance focuses attention on the precursor categories for which there is less protection. Because the analyses of these two ranking methods are quite different, the program should establish procedures and criteria for each.
3. Provide a means of feedback to the industry. Analysis is useless unless it is reflected in the design, operation, and maintenance of facilities and systems. Vulnerabilities must be addressed either to reduce the frequency of occurrence or to increase resistance to the consequences.

To accomplish these goals, an accident precursor program should have the following characteristics:

1. The program should be owned by a recognized authority in the industry and should be driven by consistent, robust goals and objectives that address the needs of the future. Operational events should be considered precursors to more serious events; from these precursors, the program should provide insights into improving safety in the future.

2. The program must be supported by an infrastructure that can sustain it. A system must be in place for gathering appropriate operational data and providing access to data providers when more detailed information is needed. Barriers to full and honest disclosure, such as proprietary information and fear of repercussions, must be addressed. Also, industry members must have incentives (either voluntary or by regulatory action) for participating.
3. The program should provide a trending and tracking system to correlate changes in industry design and practices with changes in the occurrence and nature of observed precursors. The system should also be able to distinguish between changes in trends that reflect real progress in the field and changes attributable to maturing of the process and program. The program could then provide excellent feedback to the industry on the real impact of the precursor program.
4. Systems and methods should be sensitive enough to identify an operational event as a precursor without generating too many "false detects" of events of little interest. The event-reporting requirements and event screening and selection criteria and processes must remain consistent over time to support trending and analysis.
5. Risk assessment in the industry must be mature enough to instill confidence that potential accident sequences have been identified and that the models used to assess events are sufficient and only need changes that reflect the configurations and operating practices of specific facilities. Risk models must be updated to reflect improvements in facilities, but these changes should be made in a way that does not change the level of detail or the scope of coverage. This will facilitate trending and comparison over the years.
6. Analysis should be performed on a continual basis by a consistent team of analysts to ensure the timeliness and consistency of results.

REFERENCES

Belles, R.J., J.W. Cletcher, D.A. Copinger, B.W. Dolan, J.W. Minarick, and L.N. Vanden Heuvel. 1995. Precursors to Potential Severe Core Damage Accidents: 1994, A Status Report. NUREG/CR-4674, Vols. 21–22, December 1995. Washington, D.C.: U.S. Nuclear Regulatory Commission.

Belles, R.J., J.W. Cletcher, D.A. Copinger, B.W. Dolan, J.W. Minarick, and M.D. Muhlheim. 1997a. Precursors to Potential Severe Core Damage Accidents: 1995, A Status Report. NUREG/CR-4674, Vol. 23, April 1997. Washington, D.C.: U.S. Nuclear Regulatory Commission.

Belles, R.J., J.W. Cletcher, D.A. Copinger, B.W. Dolan, J.W. Minarick, and M.D. Muhlheim. 1997b. Precursors to Potential Severe Core Damage Accidents: 1996, A Status Report. NUREG/CR-4674, Vol. 25, December 1997. Washington, D.C.: U.S. Nuclear Regulatory Commission.

Belles, R.J., J.W. Cletcher, D.A. Copinger, B.W. Dolan, J.W. Minarick, and M.D. Muhlheim. 1998. Precursors to Potential Severe Core Damage Accidents: 1997, A Status Report. NUREG/CR-4674, Vol. 26, November 1998. Washington, D.C.: U.S. Nuclear Regulatory Commission.

Copinger, D.A., and G.T. Mays. 1993. Precursors to Potential Severe Core Damage Accidents: 1992, A Status Report. NUREG/CR-4674, Vols. 17–18, December 1993. Washington, D.C.: U.S. Nuclear Regulatory Commission.

Cottrell, W.B., J.W. Minarick, P.N. Austin, E.W. Hagen, and J.D. Harris. 1984. Precursors to Potential Severe Core Damage Accidents: 1980–1981, A Status Report. NUREG/CR-3591, July 1984. Washington, D.C.: U.S. Nuclear Regulatory Commission.

Minarick, J.W., and C.A. Kukielka. 1982. Precursors to Potential Severe Core Damage Accidents: 1969–1979, A Status Report. NUREG/CR-2497, June 1982. Washington, D.C.: U.S. Nuclear Regulatory Commission.

Minarick, J.W., J.D. Harris, P.N. Austin, J.W. Cletcher, and E.W. Hagen. 1986. Precursors to Potential Severe Core Damage Accidents: 1985, A Status Report. NUREG/CR-4674, Vols. 1–2, December 1986. Washington, D.C.: U.S. Nuclear Regulatory Commission.

Minarick, J.W., J.D. Harris, P.N. Austin, J.W. Cletcher, and E.W. Hagen. 1987. Precursors to Potential Severe Core Damage Accidents: 1984, A Status Report. NUREG/CR-4674, Vols. 3–4, May 1987. Washington, D.C.: U.S. Nuclear Regulatory Commission.

Minarick, J.W., J.D. Harris, P.N. Austin, J.W. Cletcher, and E.W. Hagen. 1988. Precursors to Potential Severe Core Damage Accidents: 1986, A Status Report. NUREG/CR-4674, Vols. 5–6, May 1988. Washington, D.C.: U.S. Nuclear Regulatory Commission.

Minarick, J.W., J.D. Harris, J.W. Cletcher, P. N. Austin, and A.A. Blake. 1989. Precursors to Potential Severe Core Damage Accidents: 1987, A Status Report. NUREG/CR-4674, Vols. 7–8, July 1989. Washington, D.C.: U.S. Nuclear Regulatory Commission.

Minarick, J.W., J.W. Cletcher, and A.A. Blake. 1990a. Precursors to Potential Severe Core Damage Accidents: 1988, A Status Report. NUREG/CR-4674, Vols. 9–10, February 1990. Washington, D.C.: U.S. Nuclear Regulatory Commission.

Minarick, J.W., J.W. Cletcher, D.A. Copinger, and B.W. Dolan. 1990b. Precursors to Potential Severe Core Damage Accidents: 1989, A Status Report. NUREG/CR-4674, Vols. 11–12, August 1990. Washington, D.C.: U.S. Nuclear Regulatory Commission.

Minarick, J.W., J.W. Cletcher, D.A. Copinger, and B.W. Dolan. 1991. Precursors to Potential Severe Core Damage Accidents: 1990, A Status Report. NUREG/CR-4674, Vols. 13–14, August 1991. Washington, D.C.: U.S. Nuclear Regulatory Commission.

Minarick, J.W., J.W. Cletcher, D.A. Copinger, and B.W. Dolan. 1992. Precursors to Potential Severe Core Damage Accidents: 1991, A Status Report. NUREG/CR-4674, Vols. 15–16, September 1992. Washington, D.C.: U.S. Nuclear Regulatory Commission.

Sattison, M.B., J.A. Schroeder, K.D. Russell, S.M. Long, D.M. Rasmuson, and R.C. Robinson. 1994. SAPHIRE Models and Software for ASP Evaluations. Pp. 359–368 in Proceedings of the U.S. Nuclear Regulatory Commission's 22nd Water Reactor Safety Information Meeting. NUREG/CP-0140, Vol. 1, October 1994. Washington, D.C.: U.S. Nuclear Regulatory Commission.

USNRC (U.S. Nuclear Regulatory Commission). 1975. Reactor Safety Study: An Assessment of Accident Risks in US Commercial Nuclear Power Plants, WASH-1400. NUREG-75/014, October 1975. Washington, D.C.: U.S. Nuclear Regulatory Commission.

USNRC. 1978. Risk Assessment Review Group Report. NUREG/CR-0400, September 1978. Washington, D.C.: U.S. Nuclear Regulatory Commission.

USNRC. 1995. Systems Analysis Programs for Hands-on Integrated Reliability Evaluations (SAPHIRE) Version 5.0. NUREG/CR-6116, Vols. 1–10. Washington, D.C.: U.S. Nuclear Regulatory Commission.

Vanden Heuvel, L.N., J.W. Cletcher, D.A. Copinger, J.W. Minarick, and B.W. Dolan. 1994. Precursors to Potential Severe Core Damage Accidents: 1993, A Status Report. NUREG/CR-4674, Vols. 19–20, September 1994. Washington, D.C.: U.S. Nuclear Regulatory Commission.

Section IV
Risk Management

Inherently Safer Design

DENNIS C. HENDERSHOT
Rohm and Haas Company

> To warn of an evil is justified only if, along with the warning, there is a way of escape.
> —*Cicero*

An accident precursor can be regarded as a warning of the potential for a more serious accident, and the people responsible for the design and operation of a system must respond by identifying a "way of escape"—a risk-management strategy. The focus of this paper is on inherently safer design, that is, design that eliminates hazards, or minimizes them significantly, to reduce the potential consequences to people, the environment, property, and business. Although inherently safer design is the most robust way of addressing risk, for most facilities a complete risk-management program also includes passive, active, and procedural protections.

All systems have multiple hazards, and there can be conflicts among risks associated with different alternatives. Understanding these conflicts will enable a designer to make intelligent decisions to optimize the design. The response to an accident precursor should be similar to the response to information from any other source about hazards and risks associated with a technology. An example of how an incident-investigation team responds to an accident precursor, in this case an example from the chemical process industry, will illustrate the application of this design philosophy.

When designing or operating any engineered system, whether a chemical plant, a consumer product, a machine, or any other system, the designer must first identify specific hazards associated with the operation. Preferably, formal hazard-identification techniques are used to identify hazards as part of the design process. However, engineering design, like other human activities, is not perfect, and some hazards or specific accident scenarios that could lead to undesirable consequences will inevitably be discovered as accident precursors, and perhaps accidents, during the operation of the system. Regardless of the mechanism of discovery, the design engineer must identify a strategy for managing the hazard and its associated accident scenarios.

The concept of inherently safer design, first articulated by Trevor Kletz of ICI in 1978, has been greatly elaborated since then (CCPS, 1996; Kletz, 1978, 1998). Inherently safer design can be considered a subset of "green chemistry" and "green engineering," a more general philosophy that addresses a wide range of environmental hazards. In recent years, the chemical process industry has increasingly focused on eliminating hazards from chemical processes and plants rather than accepting their existence and designing systems to manage them; the industry has attempted to eliminate safety hazards and minimize the immediate impacts of single events, such as fires, explosions, and short-term toxic impacts. Because the strategies of inherently safer design are not specific to any industry; the example of the chemical process industry can be useful for a broad range of other technologies.

To understand the concept of inherently safer design, it is essential first to define "hazard." For purposes of this discussion, a hazard is "an inherent physical or chemical characteristic that has the potential for causing harm to people, the environment, or property" (CCPS, 1992). A hazard is intrinsic to a material or its conditions of use and, therefore, cannot be eliminated without changing the material or the conditions of use. Some examples of hazards are listed below:

- the sharp blade of a rotary lawn mower that rotates at high speed
- a cylinder of compressed air at 500 psig pressure, which contains a large amount of energy
- flammable gasoline
- the toxicity (by inhalation) of chlorine gas

Over the years, engineers have developed many tools for identifying hazards in various areas of technology. In the chemical process industry, these tools range from checklists and informal brainstorming sessions to formal, disciplined methodologies, such as hazard and operability (HAZOP) studies and failure modes and effects analysis (FMEA) (CCPS, 1992). These tools help designers understand hazards associated with the system and identify potential accident scenarios (i.e., an undesired impact on a receptor of interest, such as people, the environment, property, or business). These tools are most effective when they are used during the design process, but they can also be used to identify risk in existing systems.

The purpose of using these tools is to identify hazards and the specific accident scenarios associated with them before an accident occurs. In some cases, however, hazards or potential accident scenarios are identified through the recognition of accident precursors during operation. In all cases, designers must understand hazards and potential accident scenarios and provide adequate safeguards in the design.

RISK-MANAGEMENT STRATEGIES

Risk-management strategies can be divided into four general categories: inherent, passive, active, and procedural (CCPS, 1996):

- **Inherent** risk management involves the elimination of a hazard or the reduction of the magnitude of a hazard to the point that its consequences on potential subjects of interest are tolerable.
  —A string trimmer would eliminate the hazard of a sharp, rapidly rotating cutting blade for cutting grass. (Sheep or goats might be considered an inherently safer and more environmentally friendly technology for keeping grass trimmed. Of course, safety and environmental issues are also associated with keeping animals.)
  —Water-based paints and coatings would eliminate the fire and toxicity hazards associated with solvent-based paints.
  —A flammable, toxic extraction solvent could be replaced by supercritical carbon dioxide.

- **Passive** risk management involves devices that control or mitigate the consequences associated with a hazard without requiring sensing elements or moving parts. Passive devices function simply because they exist.
  —The deck of a rotary, power lawn mower acts as a guard. The hazard (a sharp, rapidly rotating metal blade) still exists, but the deck effectively prevents contact with hands or feet by virtue of its design and construction.
  —A reaction capable of generating 120 psig pressure from a worst-case runaway reaction is contained in a vessel designed to withstand 200 psig. The hazard (120 psig pressure) can still occur, but it is contained by the vessel, thus eliminating the need for sensors to monitor pressure or action by any device.

- **Active** risk management involves alarms, interlocks, and mitigation systems designed to detect an unsafe condition and put the system into a safe state, usually either by taking emergency action to return the system to normal operating conditions or by shutting it down. Active systems may be designed to prevent an accident or to minimize the consequences of an accident.
  —A "dead man" switch on a power lawn mower is designed to prevent an accident by disengaging the blade if the operator is not holding the mower handle.
  —A sprinkler system detects a fire and sprays it with water to minimize its spread and potential damage. This system is designed to mitigate the consequences of an accident; it does not prevent the fire, but it reduces fire damage.

—A high-level switch can prevent an accident in a storage tank by detecting an impending overfill and shutting the tank feed valve and stopping the transfer pump before the contents overflow.

- **Procedural** risk management involves standard operating procedures, operator training, safety checklists, and other management systems that depend on people.
  —The operator of a lawn mower can be trained to wear steel-toed safety shoes and safety glasses when mowing the lawn.
  — Many states require that seat belts be worn by drivers and passengers in cars.
  —A chemical plant operator can be trained to shut off reactant feeds and fully cool a reactor if the temperature exceeds 75°C.

Usually (but not always), the order of effectiveness of these strategies in terms of reliability and robustness is: inherent, passive, active, and procedural. But real systems have multiple hazards and thus require a combination of most or all of these strategies. In fact, strategies that inherently reduce one hazard may increase another hazard or even create a new hazard. Take, for example, the inherently safer design strategy of replacing a flammable, toxic extraction solvent with supercritical carbon dioxide. Supercritical extraction with carbon dioxide requires high temperature and high pressure, which would introduce new hazards to the process.

INHERENTLY SAFER DESIGN STRATEGIES

Four main strategies—minimize, moderate, substitute, and simplify—have been developed to help designers identify inherently safer systems. The **minimizing** strategy involves reducing the amount of hazardous material or energy in the system, ideally to the point that the uncontrolled release of the entire inventory of material or energy would not cause significant damage. In the chemical process industry, for example, this strategy includes considering the inventory of hazardous raw materials, in-process intermediates, and products. Other examples can be cited. The Centers for Disease Control and Prevention (CDC) reports that every day 300 young children are taken to hospital emergency rooms as a result of burns from household water that is too hot. Thus, the CDC recommends a maximum temperature of 125°F for home hot water to prevent burns (CDC, 2002). Another example involves a process that required ethylene oxide as a raw material. The ethylene oxide was shipped to the plant and stored in a large tank prior to use. To minimize the risk, a new plant was built adjacent to the ethylene oxide plant so ethylene oxide could be delivered by pipeline, thus eliminating the need for transportation and storage (Orrell and Cryan, 1987). In another case, a company in Europe used phosgene to manufacture fine chemical intermediates. The phosgene was manufactured in a separate process, and intermediate storage of many tons of toxic phosgene was required. A new, continuous process to manufacture phosgene "on demand" was developed to reduce the inventory dramatically. The new manufacturing process was continuous, essentially a "phosgene machine." When the consumer process needed phosgene, the new continuous process was started up and brought quickly to steady state to produce acceptable quality phosgene; the phosgene was then fed directly to the consumer process with no intermediate storage (Delseth, 1998; Osterwalder, 1996). In still another example, a chlorination process in a batch-stirred tank reactor was replaced by a process using a loop reactor with intensive mixing. The new reactor was one-third the size of the original, reduced batch time by 75 percent, and reduced chlorine consumption by 50 percent, to the theoretical minimum amount (CCPS, 1996).

The **substitution** strategy involves using a less hazardous reaction chemistry, or replacing a hazardous material with a less hazardous substitute. Here are some examples. In a municipal swimming pool, a solid chlorinating agent can be used instead of cylinders of chlorine gas to disinfect water. Early refrigeration systems used a variety of hazardous refrigerants, including hazardous materials, such as ammonia (toxic and flammable), light hydrocarbons (flammable), and even sulfur dioxide (toxic and corrosive). In the 1930s, chlorofluorocarbon (CFC) refrigerants were introduced to eliminate these hazards. Since the potential impact of CFCs on the environment was discovered, their use is being phased out. The challenge to engineers now is to develop refrigeration systems that eliminate fire, explosion, and toxicity hazards without causing environmental damage. In some cases, creative redesign of refrigeration systems can minimize the hazards associated with refrigerants, such as light hydrocarbons. Home refrigerators that use as little as 120 grams of isobutane refrigerant have been designed—a good example of the inherent safety strategies of minimizing and substituting.

Another example of the substitution strategy is in reaction chemistry. For many years, acrylate esters were manufactured using the Reppe process:

$$CH \equiv CH + CO + ROH \xrightarrow[HCl]{Ni(CO)_4} CH_2 = CHCO_2R$$

This process involves numerous hazards: acetylene is reactive and flammable; carbon monoxide is toxic and flammable; nickel carbonyl is toxic, an environmental hazard (heavy metal), and a suspected carcinogen; and anhydrous hydrogen chloride is toxic and corrosive. Today, most acrylate production uses a propylene oxidation process:

$$CH_2 = CHCH_3 + \frac{3}{2}O_2 \xrightarrow{Catalyst} CH_2 = CHCO_2H + H_2O$$

$$CH_2 = CHCO_2H + ROH \xrightarrow{H^+} CH_2 = CHCO_2R + H_2O$$

Although the propylene oxidation process cannot be described as inherently safe (there are hazards, primarily flammability, that must be managed), it is clearly inherently safer than the Reppe process.

The **moderating** strategy involves using a material in a less hazardous form, or under less severe conditions. Plastic materials used in molding and fabrication processes, for example, are safer if they are handled as pellets or granules rather than as fine powders, which have the potential to form explosive dust clouds. In another example, Alfred Nobel, in 1867, invented dynamite, a safer way of using nitroglycerine because it was absorbed in an inert carrier. A third example of the moderating strategy is the substitution of 28-percent aqueous ammonia solution for anhydrous (100 percent) ammonia in a neutralization application. This change reduced the downwind distance for a hazardous vapor cloud in case of a leak by a factor of up to 10, depending on weather conditions and the exact conditions of the leak.

The fourth strategy is **simplification**, which involves eliminating unnecessary complexities to reduce the likelihood of human error. A few examples follow. In 1828, Robert Stevenson, one of the pioneers of railway development, argued for simplifying controls on early steam locomotives. Stevenson recognized that complex controls made it much more likely that locomotive drivers would make mistakes that could lead to accidents. In describing existing controls, he said, "In their present complicated state they cannot be managed by 'fools,' therefore they must undergo some alteration or amendment" (Rolt, 1960). In *Turn Signals Are the Facial Expression of Automobiles*, D.A. Norman (1992) describes the poor design of a kitchen stove (Figure 1), which seems to be designed to encourage operator error.

We would prefer to blame the design of the stove on somebody in the marketing department who thought this would be a clever design that would increase sales for some reason. One would like to think that engineers designing a chemical plant would never do anything this silly. However, I actually worked in a plant where the control room and process equipment were laid out as shown in Figure 2, which exhibits essentially the same design error as the stove (Figure 1). Engineers can't blame marketing for this one!

A final example of the moderating strategy is of a company that developed a process for manufacturing methyl acetate that used reactive distillation to reduce the number of major pieces of equipment from eight columns, one extraction

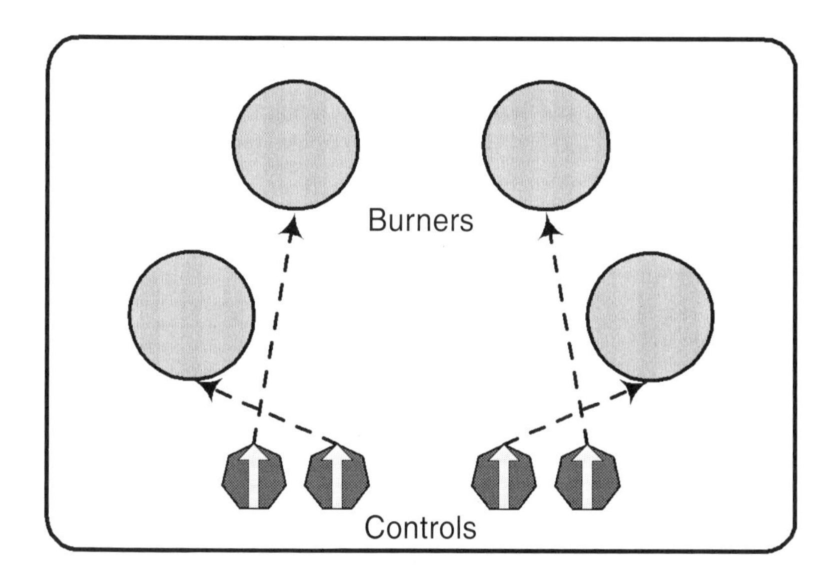

FIGURE 1 Poorly designed kitchen stove. Source: Norman, 1992.

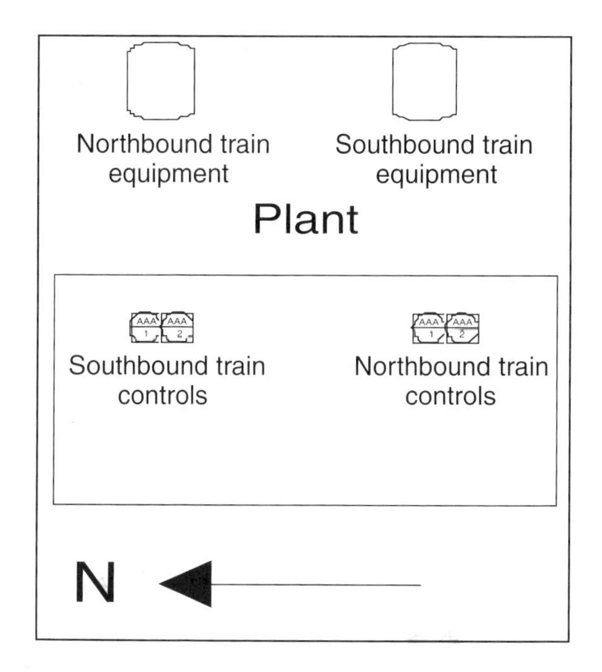

FIGURE 2 Poorly designed control room. Source: Norman, 1992.

column, and a reactor (Figure 3a) to one reactive distillation column and two additional columns (Figure 3b). The changes also eliminated all of the condensers, reboilers, instrumentation, piping, flanges, and other support equipment required for the original process (Agreda et al., 1990; Siirola, 1995).

CONFLICTS IN INHERENTLY SAFER DESIGN

A design can only be described as inherently safer in the context of a particular hazard. The design may or may not be inherently safer with respect to another hazard of the system. Therefore, design must be evaluated separately for each hazard. Thus, a water-based acrylic latex paint is inherently safer than a solvent-based paint with respect to flammability. It is also inherently safer with respect to the toxicity of the solvent (a water carrier for the latex paint). However, because of the low toxicity, latex paints may be capable of supporting the growth of microorganisms in the paint, which may make the paint useless and may also present a hazard to users. This hazard can be overcome by adding a biocide, but this is an "add-on" safety/usability feature, probably best described as "procedural" (because the manufacturer's procedures must include adding the biocide), rather than "inherent."

Designers must always remember that any modification, even one intended to improve safety, changes the system, and all changes have the potential to introduce new hazards or to increase the magnitude of existing hazards. The world is complex and interconnected. As John Muir said in *My First Summer in the Sierra* (1911), "When we try to pick out anything by itself, we find it hitched to everything else in the universe." A system designer must always be aware that any modification to a system, including the introduction of new safety features, can create or increase hazards. He or she must work to identify those hazards and make decisions based on the optimal design.

A recent example of a safety device that introduced or increased hazards is the original design of air bags in automobiles. Air bags are active safety devices. The air bag system includes one or more sensors to detect a collision, logic elements to receive the signals from the sensors and deploy the air bag, and, finally, the air bag itself. The system is intended to protect the occupants of the front seat of an automobile from injury in case of a collision. Following the large-scale commercial introduction of air bags, it was found that they sometimes caused serious injuries, even deaths, to small people (usually women and children) when the air bags were activated in collisions judged to be not severe enough to have resulted in serious injury without them. The initial response to this discovery was to recommend that children always ride in the back seat of a car equipped with air bags (a procedural response) and to allow people to disarm air bags (an inherent response that eliminated the hazard but exposed occupants to increased risk from a different hazard in the event of a serious collision). Subsequent generations of automobile air bags have been designed to be less

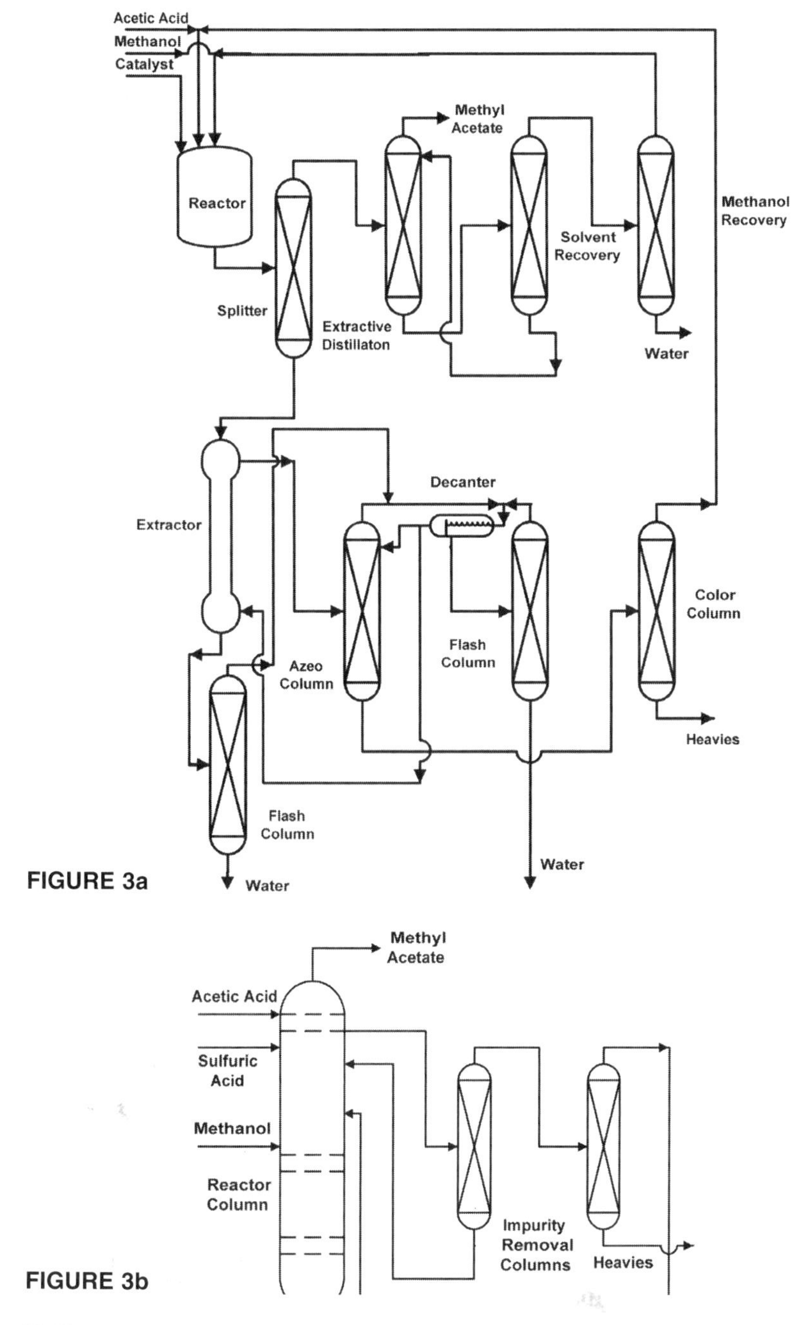

FIGURE 3 Process for manufacturing methyl acetate. **3a.** Original process. **3b.** Moderated process.

energetic, providing adequate protection in case of a serious collision but not causing serious injury in less severe impacts. This change might also be considered an example of an inherently safer design, based on the principle of moderating risk.

RESPONDING TO ACCIDENT PRECURSORS THROUGH DESIGN

Accident precursors are one way designers learn about potential hazards in consumer products, industrial machines, chemical plants, or any other engineered system. Ideally, a designer will respond to accident precursors by reevaluating the original assumptions about potential hazards and risks and redesigning the system accordingly. If the accident precursors identify new hazards or indicate a higher risk of previously known hazards, the designer should ask himself or herself a series of questions as a basis for redesigning the system.

1. Can I redesign the system to eliminate the identified hazard completely?

This is the inherently safer design approach. For a chemical process, for example, many checklists are available for specific types of equipment to help the designer identify inherently safer design strategies (CCPS, 1998). These checklists expand on the general principles of inherently safer design described above, focusing on the specific characteristics of common chemical processing equipment, such as distillation columns, reactors, and heat exchangers.

2. Can I modify the system to reduce the potential damage from the hazard?

Although the ultimate goal of inherently safer design is to eliminate hazards completely, this is not always possible. A secondary goal is to reduce the magnitude of a hazard significantly, thereby reducing the potential consequences to the receptor of concern. Checklists for inherently safer design will also help the designer identify opportunities to reduce the magnitude of the consequences of an incident. Ideally, if a hazard cannot be completely eliminated, its magnitude can be reduced to the point that it is no longer capable of causing serious injury or damage to the environment or property.

3. Do the modifications to the system identified in Questions 1 and 2 introduce new hazards or increase the potential damage from existing hazards?

Once potential system improvements have been identified, a designer must recognize that the system will be changed. While the designer was focused on improving safety, he had concentrated his efforts on a particular hazard or set of hazards. Now, the designer must step back and reevaluate the entire system, considering all hazards, using the appropriate system hazard identification tools (e.g., process safety checklists, a HAZOP, or FMEA for a chemical process). If a new hazard is identified or existing hazards are increased, the benefits of the changes must be evaluated in terms of the initial hazard and in terms of the cost of introducing new hazards or increasing the magnitude of potential damage from other existing hazards.

The designer may also evaluate the relative difficulties, cost, and effectiveness of other risk management strategies (passive, active, procedural) for managing the overall risk. This is the central problem of all engineering design. It is rarely possible to optimize a design in a way that maximizes all desirable attributes and minimizes all undesirable attributes. The true art of engineering is understanding the trade-offs and conflicts in alternative designs and selecting the one that best meets the needs of all stakeholders.

4. What passive, active, and procedural design features are necessary to manage the risk from the hazards that inevitably remain in the design?

It is unlikely that any design can eliminate all hazards in a technology. Passive, active, and procedural layers of protection are always necessary to meet safety goals. Too often, engineers accept the hazards in a system and immediately look for systems and procedures to control and manage them. A better approach is to ask first if the hazards can be eliminated or significantly reduced. Perhaps the answer is no, but it is absolutely certain that they will not be eliminated or reduced if nobody ever asks the question.

5. What general lessons can be derived from my understanding of this hazard, and how can this knowledge be applied to other systems?

Specific incidents or accident precursors are always instances of a general type of incident. Engineers often focus exclusively on the details of an occurrence and fail to recognize the general lessons that can be derived from the event. The greatest improvements to overall safety are derived from an understanding of general lessons and their broad application throughout a technology, company, or industry. This requires that the engineer step back from the specific details of the incident, identify the general technical or managerial root causes, identify other processes, equipment, or products to which these causes might also apply, and share the incident, actions, and general lessons with all interested parties. The exact circumstances that led to a specific incident are unlikely to recur in precisely the same way, even if nothing is changed. But similar incidents are likely to occur, and lessons from an event in one facility can result in improvements in other facilities that appear to have little in common with the facility where the incident occurred.

CASE STUDY

This precursor event at a chemical processing plant illustrates how the design strategies I've described can be applied. The accident precursor was the rupture of a pipe in an unoccupied building in a plant that had been operating for many years. An investigation of the incident revealed previously unknown reactive chemistry hazards, and the follow-up to the incident included significant modifications in the design of equipment, which, at first glance, did not appear to have much to do with the equipment where the incident occurred.

The plant, which had been in operation for many years, manufactured methyl isothiocyanate ($H_3CN=C=S$). The process reacts ammonium thiocyanate with methyl chloride to form methyl thiocyanate, which is isomerized to form methyl isothiocyanate. The process also involves a number of distillation and purification steps. In one part of the plant, a number of waste streams were collected for distillation to recover valuable materials for recycling before disposal of the residue as hazardous waste; some of the waste streams contained water.

The incident occurred when waste material was transferred from a collection tank to the distillation vessel through a steam-heated, insulated pipe. Following the transfer, the pipe was left full of the material, following standard operating procedures that had been in use for more than 20 years. This occurred late on a Friday night. The unit was then shut down and the building left unoccupied over the weekend. Early Saturday morning, the pipe ruptured, damaging some nearby piping and releasing a small amount of material. There were no injuries because the building was not occupied when the incident occurred. (For a more complete description, see Hendershot et al., 2003.)

Initially, it was believed that the rupture was most likely the result of hydrostatic pressure in a heated, liquid-filled line with valves closed on both ends. But other possibilities were also investigated, and subsequent laboratory work revealed that the cause was a decomposition reaction of the material in the pipe. The reaction was promoted by the presence of water in the combined waste stream and was initiated by a failure of the steam-pressure controller, resulting in maximum steam pressure and temperature on the pipe heat tracing. The decomposition reaction was previously unknown. It had taken more than 20 years for the right combination of events (sufficient water mixed with the organic components, a completely closed system, and heating to a temperature sufficient to initiate the decomposition reaction) to cause a rupture.

Actions

The investigation team did an excellent job of generalizing the lessons from the incident and identifying improvements that would contribute to the overall safety of the facility. Specifically, they asked themselves if the decomposition reaction that occurred in this pipe could occur elsewhere in the process. Laboratory experiments revealed several other areas where a similar decomposition could occur, and the team recommended improvements to eliminate the hazard, reduce its likelihood, or mitigate its effects.

Inherent Risk Management

First, the company eliminated sources of water or water-contaminated process streams to all vessels, wherever this was feasible. This was an application of the minimizing inherently safer design principle. The amount of energy from the chemical decomposition reaction was minimized by eliminating the water necessary for the reaction to occur.

Passive Risk Management

No passive design improvements were identified in this example. However, if existing vessels are replaced in the future, or if a new plant is built, it may be feasible to build stronger vessels (higher design pressure) to contain potential decomposition reactions.

Active Risk Management

Several steps were taken to manage the risk of a recurrence. First, high-temperature interlocks to shut off steam heating were provided for several vessels that could potentially have been heated to the decomposition temperature. Second, existing rupture disks or relief valves on several vessels were determined to be too small to protect the vessels from overpressurization from the newly discovered decomposition reaction. To address the problem, adequate-size relief devices were designed and installed. Third, the faulty pressure regulator on the steam tracing for the ruptured pipe was replaced with a new one, and a pressure-relief valve was installed downstream of the regulator to limit the possible pressure (and temperature) in the event of a future failure of the regulator. Finally, several vessels were provided with a dry nitrogen blanket to prevent water from humid air entering the vessels when they were emptied.

Procedural Risk Management

Standard operating procedures were modified to include draining or blowing all heated liquid lines following transfers. In addition, the preventive maintenance program for the steam pressure regulators for pipeline steam tracing was upgraded.

DESIGN STRATEGIES

We can now analyze how the design team responded to the questions for hazard and risk management.

1. Can I redesign the system to completely eliminate the identified hazard?

The design team eliminated the potential for reaction in many vessels by eliminating all sources of water.

2. Can I modify the system to reduce the potential damage from the hazard?

In places where the risk of water contamination could not be completely eliminated, the system was modified to reduce the amount of water that could get into that part of the system. Thus, the potential energy of the decomposition reaction was reduced.

3. Do the modifications to the system identified in Questions 1 and 2 introduce new hazards or increase the potential damage from existing hazards?

The investigation team recognized that process streams that contained water, which had previously been recycled, would now have to be disposed of. The team evaluated the relative costs and determined that the risk associated with the potential for a decomposition reaction outweighed the costs and risks associated with the disposal of a relatively small waste stream.

4. What passive, active, and procedural design features are required to adequately manage risk from the hazards which inevitably will remain in the design?

The investigation team identified the active and procedural changes described above to manage the residual risks.

5. What general lessons are derived from my understanding of this hazard and how can this knowledge be applied to other systems?

The investigation team applied the knowledge about the decomposition chemistry throughout the plant. In fact, most of the modifications were far removed from the pipe in which the incident occurred. Furthermore, this incident is an example of a general concern about reactive chemistry hazards, and particularly, reactive chemical structures. The investigation results were shared throughout the company, along with additional information about structures where the possibility for decomposition reactions should be evaluated. The incident was also shared with the entire industrial community through a presentation at a meeting of the American Institute of Chemical Engineers (AIChE) and subsequent publication in an AIChE journal (Hendershot et al., 2003). In fact, with the present paper, we continue to share the general lessons from this incident in the hope that we will raise awareness of reactive chemistry hazards and the importance of looking for general lessons from incident precursors throughout industry.

SUMMARY

An effective program for recognizing and responding to accident precursors must include actions taken in response to the lessons learned from the precursors. Unless action is taken, the precursor program will be simply an exercise in data collection, and no changes or improvements will be made in the system. Responding to accident precursors requires a combination of management and engineering-design activities. Accident precursors provide new information about hazards, potential accident scenarios, and the effectiveness of existing safeguards. They either reveal hazards that were previously unrecognized or provide real information about potential risks associated with hazards or accident scenarios that are already known. The engineering-design response to this information involves a combination of inherent, passive, active, and procedural strategies. In general, the most effective strategy is to eliminate, or greatly minimize, the hazard (inherently safer design). However, most real systems require a combination of layers of protection incorporating features of all four categories of response.

REFERENCES

Agreda, V.H., L.R. Partin, and W.H. Hesie. 1990. High-purity methyl acetate via reactive distillation. Chemical Engineering Progress 86(2): 40–46.

CCPS (Center for Chemical Process Safety). 1992. Guidelines for Hazard Evaluation Procedures, 2nd ed. New York: American Institute of Chemical Engineers.

CCPS. 1996. Inherently Safer Chemical Processes: A Life Cycle Approach. New York: American Institute of Chemical Engineers.

CCPS. 1998. Guidelines for Design Solutions for Process Equipment Failures. New York: American Institute of Chemical Engineers.

CDC (Centers for Disease Control and Prevention). 2002. Hot Water Burns. Available online at *http://www.cdc.gov/nasd/docs/d000701-d000800/d000702/d000702.html.*

Delseth, R. 1998. Production industrielle avec le phosgene. Chemia 52(12): 698–701.

Hendershot, D.C., A.G. Keiter, J. Kacmar, J.W. Magee, P.C. Morton, and W. Duncan. 2003. Connections: how a pipe failure resulted in resizing vessel emergency relief systems. Process Safety Progress 22(1): 48–56.

Kletz, T.A. 1978. What you don't have, can't leak. Chemistry and Industry. May 6 (No. 9): 287–292.

Kletz, T.A. 1998. Process Plants: A Handbook for Inherently Safer Design. Philadelphia, Pa.: Taylor and Francis.

Muir, J. 1911. My First Summer in the Sierra. Boston: Houghton Mifflin.

Norman, D.A. 1992. Turn Signals Are the Facial Expressions of Automobiles. Reading, Mass.: Addison-Wesley.

Orrel, W., and J. Cryan. 1987. Getting rid of the hazard. The Chemical Engineer, August 1987, pp. 14–15.

Osterwalder, U. 1996. Continuous Process to Fit Batch Operation: Safe Phosgene Production on Demand. Pp. 6.1–6.6 in Symposium Papers, Institute of Chemical Engineers, Northwest Branch. Rugby, U.K.: Institute of Chemical Engineers.

Rolt, L.T.C. 1960. The Railway Revolution: George and Robert Stevenson. New York: St. Martin's Press.

Siirola, J.J. 1995. An industrial perspective on process synthesis. American Institute of Chemical Engineers Symposium Series 91: 222–223.

Checking for Biases in Incident Reporting

TJERK VAN DER SCHAAF and LISETTE KANSE
Department of Technology Management
Eindhoven University of Technology

Incident reporting schemes have long been part of organizational safety-management programs, especially in sectors like civil aviation, the chemical process industry, and, more recently, rail transport and in a few health care domains, such as anaesthesiology, pharmacies, and transfusion medicine. In this paper, we define incidents as all safety-related events, including accidents (with negative outcomes, such as damage and injury), near misses (situations in which accidents could have happened if there had been no timely and effective recovery), and dangerous situations.

But do reporting schemes capture a representative sample of actual events? One of the reasons incident-reporting databases might be biased is a tendency to over- or underreport certain types of events. To address the vulnerabilities of voluntary reporting schemes in terms of the quantity and quality of incident reports, guidelines have been developed for designing and implementing such schemes. Reason (1997) lists five important factors for "engineering a reporting culture": (1) indemnity against disciplinary proceedings; (2) confidentiality or de-identification; (3) separation of the agency that collects and analyzes the reports from the regulatory authority; (4) rapid, useful, accessible, and intelligible feedback to the reporting community; and (5) the ease of making the report. Another expert, D.A. Lucas (1991) identifies four organizational factors: (1) the nature of the information collected (e.g., descriptive only, or descriptive and causal); (2) the uses of information in the database (e.g., feedback, statistics, and error-reduction strategies); (3) analysis aids to collect and analyze data; (4) organization of the scheme (e.g., centralized or local, mandatory or voluntary). Lucas also stresses the importance of the organization's model of why humans make mistakes, as part of the overall safety culture.

These are just two examples of organizational design perspectives on reporting schemes. Much less is known about the individual reporter's perspective: (1) when and why is an individual inclined to submit a formal report of a work-related incident; and (2) what aspects of an incident is an individual able and willing to report.

The starting point for the investigation described in this paper was an observation made during a reanalysis of part (n = 50 reports) of a large database of voluntarily reported incidents at a chemical process plant in the Netherlands, where we encountered very few reports of self-made errors (Kanse et al., in press). This was surprising because this plant had been highly successful in establishing a reporting culture; minor damage, dangerous situations, and large numbers of near misses (i.e., initial errors and their subsequent successful recoveries) were freely reported, two reports per day on average from the entire plant. The 200 employees of the plant, as well as temporary contract workers, contributed to the plant's near-miss reporting system (NMRS), which had been operational for about seven years by the time we performed the reanalysis. The NMRS was regarded as a "safe" system in terms of guaranteed freedom from punishment as a result of reporting an incident. Even more puzzling was that references to self-made errors were also absent in the particular subset we were analyzing—successfully recovered (initial) errors (human failures) and other failures, which were thus completely inconsequential. The question was why plant operators did not report successful recoveries from self-made errors.

To address this question, we began by reviewing the literature on the reasons individuals fail to report incidents in general and then evaluating their relevance for our study. We then generated a taxonomy of possible reasons for nonreporting. Next, we instituted a diary study in which plant operators were asked to report their recoveries from self-made errors under strictly confidential conditions, outside of the normal NMRS used at the plant. In addition to descriptions of recovery events, we asked them to indicate whether or not (and *why*) they would normally have reported the event. We then categorized the reasons according to our taxonomy. The results are discussed in terms of the reporting biases we identified and possible countermeasures to improve the existing reporting system.

REASONS FOR NOT REPORTING

We began our search with the Psychinfo and Ergonomics Abstracts databases to ensure that we covered both the domain of work and organizational psychology and the domain of ergonomics, human reliability, and safety. The key words we used were "reporting system and evaluation," "reporting barriers," "reporting tendencies," "reporting behavior," "reporting biases," "incident report," "near miss report," and, in Psychinfo, simply "near miss." We included truncated forms of the keywords (i.e., "report*" for report, reporting, and reports) and alternative spelling to maximize the scope of our search. We assessed potential relevance based on the abstracts; we also added references to our review from the reference lists of items we had selected.

Based on our search, we concluded that, even though there is a relatively large body of literature on *organizational design* guidelines for setting up incident reporting schemes, very few insights could be found into the reasons *individuals* decided whether or not to report an incident. We grouped the factors influencing incident reporting into four groups:

- *fear* of disciplinary action (as a result of a "blame culture" in which individuals who make errors are punished) or of other people's reactions (embarrassment)
- *uselessness* (perceived attitudes that management would take no notice and was not likely to do anything about the problem)
- *acceptance of risk* (incidents are part of the job and cannot be prevented; or a "macho" perspective of "it won't happen to me")
- *practical reasons* (too time consuming or difficult to submit a report)

Adams and Hartwell (1977) mention the blame culture (as does Webb et al., 1989) and the more practical reasons of time and effort (see also Glendon, 1991). Beale et al. (1994) conclude that the perceived attitudes of management greatly influence reporting levels (see also Lucas, 1991, and Clarke, 1998) and that certain kinds of incidents are accepted as the norm. Similarly, Powell et al. (1971) find that many incidents are considered "part of the job" and cannot be prevented. This last point is supported by Cox and Cox (1991), who also stress the belief in personal immunity ("accidents won't happen to me"; see also the "macho" culture in construction found by Glendon, 1991). O'Leary (1995) discusses several factors that might influence a flight crew's acceptance of the organization's safety culture, and thus the willingness to contribute to a reporting program. These factors include a lack of trust in management because of industrial disputes; legal judgments that ignore performance-reducing circumstances; pressure from society to allocate blame and punish someone for mistakes; the military culture in aviation; and the fact that pilots, justifiably or not, feel responsible or even guilty for mishaps as a result of their internal locus of control combined with their high scores on self-reliance scales. Elwell (1995) suggests that the reasons human errors are underreported in aviation are that flight crew members may be too embarrassed to report their mistakes or that they expect to be punished (see also Adams and Hartwell, 1977, and Webb et al., 1989); and if an error has not been observed by others, they are less likely to report it. Smith et al. (2001) report clear differences (and thus biases in the recording system) between recorded industrial injury events and self-reported events collected via interviews and specifically developed questionnaires.

A number of publications in the health care domain are focusing increasing attention on the reporting of adverse events (i.e., events with observable negative outcomes). For example, Lawton and Parker (2002) studied the likelihood of adverse events being reported by health care professionals and found that reporting is more common in places where protocols are in place and are not adhered to than where there are no protocols in place; in addition reports are more likely when patients were harmed; near misses, they found, are likely to go unreported. The suggested explanations for a reluctance or unwillingness to report are the culture of medicine, the emphasis on blame, and the threat of litigation.

Probably the most comprehensive study so far, and to our knowledge the only one in which individuals were asked to indicate their reasons for not reporting, was undertaken by Sharon Clarke (1998). She asked train drivers to indicate the likelihood that they would report a standard set of 12 realistic incidents (a mix of dangerous situations, equipment failures, and other people's errors). The drivers were offered a predefined set of six reasons for not reporting in each case: (1) one would tell a colleague directly instead of reporting the incident; (2) this type of incident is just part of the job; (3) one would want to avoid getting someone else in trouble; (4) nothing would be done about this type of incident; (5) reporting involves too much paperwork; and (6) managers would take no notice.

How could we use the results of our search to generate a set of reasons individuals might not report recoveries from self-made errors? Using the four categories of reasons reported in the literature (fear; uselessness; acceptance of risk; practical reasons) as a starting point, we discussed the question with people from three groups of employees at the chemical plant: management, safety department, and operators. Their opinions on possible reasons for not reporting are summarized below:

- The chemical plant operators, part of a high-reliability organization (HRO), or at least something close to that (for a description of the characteristics of an HRO, see Roberts and Bea, 2001), rarely gave the reasons we found in the literature—namely the acceptance of incidents as part of their jobs or as unavoidable or a conviction that they were bound to happen.
- None of the groups said plant management systematically ignored reported risks, which could make coming forward with information useless.
- Most of the employees thought operators might be afraid or ashamed to report their own initial errors that required recovery actions.
- Employees also considered it less important to report incidents that were indicative of well known risks because they were widely known by their colleagues and, therefore, had minimal learning potential.
- Some types of incidents might not be considered appropriate to the goals of the reporting scheme.

- Another suggestion was that if they themselves could "take care" of the situation, reporting the incident would be superfluous.
- If there were ultimately no consequences, the incident could be considered unimportant.
- Finally, the lack of time ("always busy") could be a factor, as could other practical reasons (e.g., not fully familiar with the system).

Based on the considerations listed above, we propose the following six possible reasons for not reporting recoveries from self-made errors: (1) afraid / ashamed; (2) no lessons to be learned from the event; (3) event not appropriate for reporting; (4) a full recovery was made, so no need to report the event; (5) no remaining consequences from the event; (6) other factors.

THE DIARY STUDY

Methodology

Following the methods used in previous studies, we used personal diaries to get reports of everyday errors (Reason and Lucas, 1984; Reason and Mycielska, 1982; and especially Sellen, 1994). We asked all of the employees on one of the five shifts at the chemical plant to participate in a diary study for 15 working days (five afternoon shifts, five night shifts, and five morning shifts). Twenty-one of the 24 operators filled out a small form for every instance when they realized that they had recovered from a self-made error. The form contained several items: describe the self-made error(s); describe the potential consequences; tell who discovered the error(s), including how and when; describe the recovery action(s) taken; describe remaining actual consequences; and finally, "Would you have reported such an incident to the existing Near Miss Reporting System (choose from yes/no/maybe)" and "Why (especially if the answer is no)." We did not offer any preselected possible reasons as options for the last question, because we wanted the operators to feel free to express themselves.

Results

During the 15 days of the diary study, the 21 operators completed forms relating to 33 recoveries from self-made errors. In only three cases did they indicate that the incident would also have been reported to the existing NMRS; for five of the remaining cases, no reason(s) were given for not reporting.

Transcribed answers to the last question in the 25 remaining cases were given to two independent coders—the authors and another human-factors expert with experience in human-error analysis. One coder identified 32 reasons; the other found 34 reasons. The coders then reached a consensus on 32 identifiable reasons. The two coders independently classified each of the 32 reasons into one

TABLE 1 Examples of Coded Transcripts

Code Assigned	Example from Transcript
No lessons to be learned from the event	The unclear/confusing situation is already known.
Not appropriate for reporting	System is not meant for reporting this kind of event.
Full recovery made	I made the mistake and recovery myself.
No remaining consequences	Mistake had no consequences.
Other	Not reported at the time, too busy.

of the six categories. They agreed on 28 of the 32 reasons and easily reached consensus on reasons they had coded differently. A typical example of the statements and the resulting code are shown in Table 1. The overall results are shown in Figure 1.

In addition to the results shown above, the operators, on average, judged the potential consequences of the incidents in the diary study, if they not been recovered from, to be as serious as the consequences of incidents normally reported to the existing NMRS (Kanse et al., 2004). Potential consequences for each reported event were: production/quality loss, delay, damage; injury/health effects; and environmental effects. The severity for each type of consequence was also indicated (no consequences, minor consequences, considerable consequences, or major consequences).

The *remaining* consequences after recovery, however, differed from the consequences indicated for the events studied in the re-analyzed part of the existing NMRS database (Kanse et al., 2004). After recovery from self-made errors as reported in the diary study, in three events a minor delay remained, in one event minor production/quality loss remained, and one event involved minor repair costs. In contrast, there were remaining consequences in a much higher percentage of the 50 events from the NMRS (involving multiple, different types of failures per event): in six events a minor delay remained; in one event there were minor health-related consequences; in four events there were minor environmental consequences; in 14 events the hazard continued to exist for a significant time before the final correction was implemented; and in 20 events there were minor repair costs. These findings suggest that a complete and successful recovery from self-made errors may be easier to achieve than from other types of failures or combinations of failures. The main differences were in repair costs and the time during which the hazard continued.

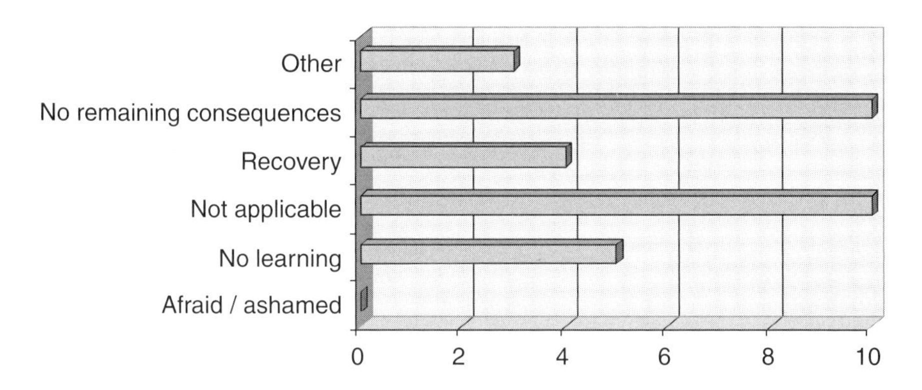

FIGURE 1 Distribution of 32 reasons given by 21 operators for *not* reporting 25 "diary incidents" to the existing NMRS.

CONCLUSIONS

In terms of the trustworthiness of results, the diary study successfully complemented and provided a check on the existing near-miss database. Respondents were open and frank with the author who collected the data, which they otherwise would not have shared with the plant management and safety staff. They also described their reasons for not reporting clearly. The fact that two independent coders were able to use the taxonomy of reasons indicates its potential usefulness for the future.

The plant's management and safety staff were somewhat surprised at the results shown in Figure 1. Some of them had expected that there would still be some fear or shame about reporting self-errors and/or a low level of perceived *potential* consequences as the major reasons successful recoveries were not reported. Thus, the results showed a genuine difference between operators and management in *perceived importance*, as measured by the options of no lessons to be learned, not appropriate for the system, full recovery, and no remaining consequences. Our hope is that the plant will now set up a program that clearly communicates management's sincere interest in learning about the personal and system factors that make successful recoveries possible and that they will not adopt an attitude of "all's well that ends well"; according to Kanse et al. (2004), the latter attitude is not compatible with the way an HRO should function. The fact that none of the participants mentioned being afraid or ashamed to report errors may be a very positive indicator of the plant's safety culture.

The success of this limited (in time and resources) diary study suggests that the procedure could be repeated after the implementation of a program to convince operators of the importance of reporting recoveries, especially successful recoveries. A follow-up study could measure the change in operators' perceptions. The second study (and subsequent studies from time to time) could monitor the emergence of other, possibly new, reasons for not reporting.

REFERENCES

Adams, N.L., and N.M. Hartwell. 1977. Accident reporting systems: a basic problem area in industrial society. Journal of Occupational Psychology 50: 285–298.

Beale, D., P. Leather, and T. Cox. 1994. The Role of the Reporting of Violent Incidents in Tackling Workplace Violence. Pp. 138–151 in Proceedings of the 4th Conference on Safety and Well-Being. Leicestershire, U.K.: Loughborough University Press.

Clarke, S. 1998. Safety culture on the UK railway network. Work and Stress 12(1): 6–16.

Cox, S., and T. Cox. 1991. The structure of employee attitudes to safety: a European example. Work and Stress 5(2): 93–106.

Elwell, R.S. 1995. Self-Report Means Under-Report? Pp. 129–136 in Applications of Psychology to the Aviation System, N. McDonald, N. Johnston, and R. Fuller, eds. Aldershot, U.K.: Avebury Aviation, Ashgate Publishing Ltd.

Glendon, A.I. 1991. Accident data analysis. Journal of Health and Safety 7: 5–24.

Kanse, L., T.W. van der Schaaf, and C.G. Rutte. 2004. A failure has occurred: now what? Internal Report, Eindhoven University of Technology.

Lawton, R., and D. Parker. 2002. Barriers to incident reporting in a healthcare system. Quality and Safety in Health Care 11: 15–18.

Lucas, D.A. 1991. Organisational Aspects of Near Miss Reporting. Pp.127–136 in Near Miss Reporting as a Safety Tool, T.W. van der Schaaf, D.A. Lucas, and A.R. Hale, eds. Oxford, U.K.: Butterworth-Heinemann Ltd.

O'Leary, M.J. 1995. Too Bad We Have to Have Confidential Reporting Programmes!: Some Observations on Safety Culture. Pp. 123–128 in Applications of Psychology to the Aviation System, N. McDonald, N. Johnston, and R. Fuller, eds. Aldershot, U.K.: Avebury Aviation, Ashgate Publishing Ltd.

Powell, P. I., M. Hale, J. Martin, and M. Simon. 1971. 2000 Accidents: A Shop Floor Study of Their Causes. Report no. 21. London: National Institute of Industrial Psychology.

Reason, J. 1997. Managing the Risk of Organisational Accidents. Hampshire, U.K.: Ashgate Publishing Ltd.

Reason, J., and D. Lucas. 1984. Using Cognitive Diaries to Investigate Naturally Occurring Memory Blocks. Pp. 53–70 in Everyday Memory, Actions and Absent-Mindedness, J.E. Harris and P.E. Morris, eds. London: Academic Press.

Reason, J., and K. Mycielska. 1982. Absent-minded?: The Psychology of Mental Lapses and Everyday Errors. Englewood Cliffs, N.J.: Prentice Hall, Inc.

Roberts, K.H., and R.G. Bea. 2001. Must accidents happen?: lessons from high-reliability organisations. Academy of Management Executive 15(3): 70–77.

Sellen, A.J. 1994. Detection of everyday errors. Applied Psychology: An International Review 43(4): 475–498.

Smith, C.S., G.S. Silverman, T.M. Heckert, M.H. Brodke, B.E. Hayes, M.K. Silverman, and L.K. Mattimore. 2001. A comprehensive method for the assessment of industrial injury events. Journal of Prevention and Intervention in the Community 22(1): 5–20.

Webb, G.R., S. Redmand, C. Wilkinson, and R.W. Sanson-Fisher. 1989. Filtering effects in reporting work injuries. Accident Analysis and Prevention 21: 115–123.

Knowledge Management in High-Hazard Industries

Accident Precursors as Practice

JOHN S. CARROLL
Sloan School of Management
Massachusetts Institute of Technology

Accident precursors are events that must occur for an accident to happen in a given scenario, but that have not resulted in an accident so far. High-hazard industries, such as nuclear power and aviation, that would put many people at risk in the event of a single accident are particularly sensitive to precursors and consider them opportunities to avoid accidents. Accidents happen when precursors occur in combination and/or when system defenses fail to mitigate a situation. Every precursor event is, therefore, both a test of the adequacy of system defenses and an opportunity to develop and apply knowledge to avoid accidents. Failure to take notice of these "tests" and to build a strong knowledge-management system is a sign of trouble ahead.

At the Three Mile Island (TMI) nuclear power plant, for example, a combination of events—a stuck-open pressure-relief valve that allowed water levels in the reactor to drop, thus uncovering the radioactive core plus indicators that showed the position of the switch controlling the valve but not the valve itself plus operator training that cautioned operators about overfilling the reactor with water—destroyed a billion dollar unit of the plant and changed the nuclear power industry forever. Even though information that could have prevented the TMI event was available from similar incidents at other plants, recurrent problems with the same equipment at TMI, and critiques of operator training, that information was not incorporated into plant-wide or industry-wide operating practices (Marcus et al., 1989). The president of the utility, Herman Dieckamp later reflected on the incident (Kemeny et al., 1979):

> To me that is probably one of the most significant learnings of the whole accident, the degree to which the inadequacies of that experience feedback loop . . . significantly contributed to making us and the plant vulnerable to this accident.

In response to the TMI accident, the nuclear power industry created the Institute of Nuclear Power Operations (INPO) to identify precursors, disseminate lessons learned and best practices, and generally ensure that every plant operates with the best knowledge available (and also to forestall further regulation). The World Association of Nuclear Operators performs these tasks globally. Although knowledge development and dissemination have been successful overall, problems continue in this industry, which is under continuous scrutiny by regulators and a wary public.

ACCIDENT PRECURSORS AND KNOWLEDGE MANAGEMENT

From a knowledge-management perspective, precursors are signals of possible problems, chinks in an operation's armor, or pathways to accidents. They are called precursors rather than accidents because systems have multiple layers of defense like slices of Swiss cheese stacked together (Reason, 1997). A precursor problem may pass through one or two layers of defense (through the holes in the Swiss cheese), but another layer usually stops the progression toward an accident. Only when "all of the holes line up" does the problem overcome or bypass all defenses and become an accident. As signals, precursors allow us to find the sources of potential problems and assess the robustness of defenses. Based on information from precursors, we can improve defenses or make sure they function as designed and add new defenses when problems become frequent or serious or new problems appear.

The history of the nuclear power industry shows a constant tension between wariness and complacency. Early on, operators and regulators believed that nuclear power would be a simple technology to operate, that electricity would be "too cheap to meter," and that safety would be assured. TMI was a "fundamental surprise" (Lanir, 1986) that caused intense scrutiny and huge investments in safety equipment, procedures, training, reporting, and people. Probabilistic risk analysis was invented as a way of anticipating problems and designing defenses against them. However, each time the industry has thought its was secure in its ability to anticipate problems and design defenses, new, unanticipated challenges have arisen, such as shut-down risk, stress corrosion cracking, and inadequacies in safety culture. The industry continues to learn, forget, and relearn a difficult lesson—that anticipation must be combined with resilience in responding to precursors (Marcus and Nichols, 1999; Weick et al., 1999; Wildavsky, 1988).

One institutionalized approach to combating problems and remaining alert is self-assessment embedded in corrective-action programs. In a speech to the Regulatory Information Conference in 1996, Dr. Shirley Jackson, former chair of the U.S. Nuclear Regulatory Commission (USNRC) attributed improvement in the 1990s to "increased emphasis by both the [US] NRC and the industry in the following three areas: (1) improved maintenance practices; (2) consideration of risk in the operation and maintenance of nuclear plants; and (3) self-assessment of events to identify root causes of problems and ensure effective corrective actions." She went on to say that self-assessment "should be an ingrained part of every licensee's way of doing business" and that self-assessment would become increasingly important as the industry moved "to more performance-oriented regulatory approaches."

THE STOCK-AND-FLOW MODEL OF KNOWLEDGE MANAGEMENT

Traditional knowledge management is a combination of maintaining repositories of explicit information and expert know-how organized by professional discipline. Examples of explicit information include databases, procedural manuals, drawings, and planning documents. Routine operations are guided by this codified knowledge, and routine problems can often be addressed by consulting the manuals. Thus, some knowledge can be explicitly codified in these reservoirs (Argote and Ingram, 2000), but some knowledge is tacit, implicit in the experience and training of individuals. Thus, engineers, operators, craftsmen, accountants, and others with expertise in particular domains have developed "judgment" and recognition-based diagnostic and action skills (Klein, 1998). Most exceptions and problems can be categorized and referred to subject-matter experts for resolution.

In this model of knowledge management, the key issue is "where" the knowledge resides. Knowledge is a stock or supply that has to be accessible and can be moved around as needed, like supplies in a warehouse or money in a bank account. When a precursor is noticed, a search is made for relevant information to ensure that defenses are adequate or to strengthen defenses if necessary. The search focuses on the problem (e.g., if the problem has been seen before, if other plants in the industry have seen it) and on the domains of expertise relevant to the problem (e.g., maintenance, engineering, chemistry). Investigators have access to databases created by a plant, groups of similar plants, manufacturers, industry groups, and even regulators. Explicit knowledge in the databases can be applied directly, and deviations are dealt with by evolutionary enhancements, including adding controls: "Safe operating procedures . . . are continually being amended to prohibit actions that have been implicated in some recent accident or incident" (Reason, 1997).

However, most problems involve knowledge that is local and contextual, tacit as well as explicit. Therefore, additional knowledge is necessary before what is known can be applied to new instances. In other words, problems may not be identical from place to place or time to time, and information may be "sticky" or difficult to move from one location to another (Szulanski, 1996; von Hippel, 1994). Expert judgment may be necessary to draw analogies, tailor solutions for particular situations, and so forth. In such cases, success depends upon the personal involvement of knowledgeable individuals and personal networks that connect accountable investigators with knowledgeable experts.

Industries such as nuclear power recognize the importance of personal contacts in the dissemination of best practices, experience with precursors, and so forth. Virtually all bits of new information include contact information for individuals who are the best sources of information. Thus, the article or the database entry is an advertisement or infomercial rather than a source of necessary information. To implement a best practice, one must learn by telephone, by visiting the source plant, by hosting peer-assist visits from source-plant personnel, or by using consultants as transmission channels. Contacts may be facilitated by liaisons, job rotations, or temporary exchanges of personnel with other plants or industry organizations, such as INPO. Thus, knowledge management depends upon the development of informal (often invisible) networks of personal contacts within a plant, with other plants, with suppliers, consultants, regulators, universities, etc. One of the first cultural precursors to trouble is an organization that withdraws from "nonessential" industry activities and, therefore, limits its access to new information and knowledgeable peers; this is what Millstone Station did in the 1980s following the financial challenges of building a third unit (Carroll and Hatakenaka, 2001).

Hansen (1999) showed that different kinds of network ties or interpersonal relationships are necessary for different kinds of knowledge transfer. Having a large number of "weak ties," that is, infrequent, distant relationships and acquaintanceships, facilitates the search for new knowledge. A person with a broad network can find new information easily, including by using e-mail and web searches. If the information is relatively simple and easy to transfer, weak ties are very efficient and useful. However, weak ties can actually slow down the transmission of complex information, which requires a strong connection among individuals or groups.

THE CAPABILITIES MODEL OF KNOWLEDGE MANAGEMENT

We can conceptualize knowledge management as a system capable of attending to signals, generating new knowledge (updating), retaining knowledge, and applying knowledge where it is needed. This constellation of capabilities is sometimes called organizational learning (Carroll et al., 2002; Crossan et al., 1999; Senge, 1990). For our purposes, organizational learning is another description of how knowledge is generated and applied in action, which includes capabilities for attending, making sense, and implementing change.

Attention or "heedfulness" is a crucial first step in reacting to precursors (Marcus and Nichols, 1999; Weick et al., 1999). In most organizations, precursors either go unheeded or are responded to at the local level with no signal reaching beyond the immediate work context. Reporting systems are an institutionalized form of attention; planning, typically understood as a way of allocating resources and controlling activities, enables people within an organization to notice things more easily and to get more rapid and more useful feedback about how things are going (deGeus, 1988).

Organizations rarely succeed because they "meet plan," but organizations without a clear plan find it hard to notice when things are not going well and, therefore, to respond to incipient problems creatively and effectively. For precursors to be recognized as precursors, there has to be a shared understanding of what is normal and what is off-normal, what is expected and what is unexpected, what is desirable and what is undesirable. As Weick and Sutcliffe (2001) state, "to move toward high reliability is to enlarge what people monitor, expect, and fear." A typical nuclear power plant, for example, formally identifies more than 2,000 problems or incidents per year, 90 percent of which would have been ignored a decade ago.

Once precursors or troublesome conditions have been noticed, some type of analysis or investigation follows. Nearly all high-hazard organizations conduct investigations of problems as part of their corrective-action programs, which start with the reporting of problems and continue with the investigation of facts and opinions, the attribution of causes, the generation of insights and recommendations, the implementation of interventions to improve performance, and the verification that these interventions are carried out and produce the expected results (Carroll, 1995, 1998; Carroll et al., 2001; van der Schaaf et al., 1991). More frequent than the massive investigations triggered by rare accidents, such as TMI, these smaller scale self-analyses and problem-solving activities focus on small defects, near misses, and other lesser failures (Sitkin, 1992) or precursors (Reason, 1990). Problem investigation is a kind of off-line, reflective practice that involves sense-making, analysis, and imagining alternatives. This often takes place outside of the regular work process, often by individuals who are not immediately involved in the problem (Argyris, 1996; Rudolph et al., 2001).

Although individuals can investigate most problems, the most serious, persistent, causally ambiguous, and organizationally complex problems are investigated by teams. Each year, nuclear power plants assemble multidisciplinary teams (sometimes including personnel from other plants, headquarters, other companies, and elsewhere) to investigate a small number of problems that seem to extend beyond the knowledge base of any single department. These teams not only provide a wide range of expertise, they also have better access to information from informants and more credibility with the audiences who must support the implementation of their recommendations. Serving on these teams can provide valuable experience and enhance an individual's knowledge and skills, which are then brought back to coworkers when the team member returns to his or her home department (Gruenfeld et al., 2000); the experience helps bridge the gap between communities of practice, thus enhancing the capabilities of the organization as a whole (Cook and Brown, 1999).

Investigations often focus on fixing immediate problems so operations can return to normal and everyone can regain a sense of predictability and control, which are so important to managers and engineers, especially in high-hazard industries (Carroll, 1998; Carroll et al., 2002). However, just as exploiting readily available information may keep one from exploring new possibilities (March, 1991), fixing immediate problems may interfere with the extraction of all useful information from a precursor event. For example, in the chemical plant pipe failure reported by Hendershot et al. (2003) or the chemical plant charge-heater fire reported by Carroll et al. (2002), investigations could have stopped with simple explanations and fixes that would have prevented those particular problems from recurring. In both cases, however, the analyses went further to identify "root causes," which resulted in new knowledge about the technology and organization of the work.

In the charge-heater fire investigation, for example, the team noted as a "Key Learning" that the plant staff had made decisions without questioning their assumptions. First, the maintenance department had changed decoking processes but did not know and never checked to be sure that the new process was effective. Second, operators increased the burner pressure in the charge heater to increase production but did not know the consequences of doing so. Third, operators changed the pattern of firing heater tubes (to fire hotter around the perimeter without setting off safety alarms) but again did not know the consequences of doing so. The investigation team found that the fire was caused by a combination of (1) operators firing heater tubes in such a way that the hottest temperatures were located away from the instruments designed to detect danger and (2) the presence of residual coke (coal dust) on the inside of the tubes that the new maintenance process had left behind. On the basis of these findings, the first recommendation for future action was that the plant identify "side effects" and be more aware of the broad "decision context" when changing production processes. This resulted in the implementation of a new "management of change" process so that the global implications of proposed local actions could be anticipated better.

THE PRACTICE MODEL OF KNOWLEDGE MANAGEMENT

Neither the stock-and-flow model nor the capabilities model describes in detail how knowledge management is accomplished. The assumption is that the right tools, people, and environment will promote the development, transfer, and use of knowledge. The practice model of knowledge management focuses on specific activities (Bourdieu, 1977; Brown and Duguid, 1991; Carlile, 2002). For example, knowledge is often embodied in stories and transmitted through storytelling. In addition, knowledge development among communities-of-practice requires specific boundary-spanning or bridging practices. Incident investigations and analyses of root causes (which include a variety of techniques for looking beyond immediate or proximal causes) may be valuable not only as analytical tools, but also as opportunities for conversations with shared purposes (Carroll et al., 2002).

In our research on incident investigation teams in nuclear power plants, we assumed that teams that used root-cause analysis to make deeper investigations of precursor events would generate more knowledge and that organizations would implement more effective changes that would improve performance. We discovered, however, that the investigation teams and the managers to whom they reported had very different ideas about what constituted a good investigation and a good report. The teams wanted to find the causes of precursors, to dig deeply and identify failed defenses. The managers wanted actionable recommendations that would fix problems and reestablish control. Managers seeking efficiency delegated participation on the team and waited to respond to a draft report rather than taking the time to work directly with the team (Nutt, 1999). As a result, the hand-off from team to manager was often ineffective. Reports were sometimes modified or negotiated to obtain manager "sign off," and recommendations were sometimes watered down or folded into other activities, or even refused, on the basis of cost or other practicalities. Managers often thought investigation teams were unrealistic, whereas the teams thought managers were defensive.

Interestingly, at the chemical company that investigated the charge-heater fire, the investigation team had an explicit goal of *educating managers*, rather than solving problems! In this company, teams presented facts and carefully reasoned causal connections, but did not make recommendations. The managers' collective job was to understand the problem and its context, discuss opportunities for improvement, commission activities to develop solutions, and implement changes.

Problem investigations provide precisely the kind of opportunities that can bring together diverse perspectives and facilitate learning and change. The mixing of occupational and educational backgrounds (Dougherty, 1992; Rochlin and von Meier, 1994) and cognitive styles (Jackson, 1992; White, 1984) that combine abstract, systemic issues with concrete, operational details and technical complexity with human ambiguity can lead to informational diversity (Jehn et al., 1999) or "conceptual slack" (Schulman, 1993). Weick et al. (1999) similarly argue that consistent reliability requires that problems not be oversimplified, which requires diverse perspectives and frequent boundary-spanning activities. This creates skills and opportunities for engaging in a process of knowing that can bring to the surface previously unarticulated mental models of the work environment, compare them, and lead to new, shared models (Cook and Brown, 1999).

In the cases we studied, boundary spanning was only partially successful. Sharpening and bridging differences among disciplines and hierarchical levels requires an atmosphere of mutual respect and trust. Managers, however, were often not full participants on the investigative teams, reports were rather casual in connecting causes and recommended actions, and negotiations over the final report tended to be about authority rather than reasoning. It takes mindful attention to build shared understanding around diffuse issues, such as "culture" and "accountability," that have very different meanings and implications to different professional groups (Carroll, 1998; Carroll et al., 2002). Because the emphasis is usually on controlling deviations from existing procedures and rules, few teams and managers are willing or able to work hard to clarify meaning and build shared mental models. Therefore, they often miss opportunities to deepen their understanding that could lead to organizational learning and change.

SUMMARY

All politics is said to be local, and in important ways knowledge is local as well. In managing knowledge about accident precursors, organizations must attend to the local nature of problems and the knowledge that must be brought to bear to address them, as well as to the global nature of what is learned and what may be needed at other times in other locations. In addition, they must consider knowledge not only as a stock of information, but also as providing the capability of inquiring, imagining, bridging boundaries, building networks of trusting relationships, and taking action. Precursor events are opportunities to enact and improve organizational practices.

REFERENCES

Argote, L., and P. Ingram. 2000. Knowledge transfer: a basis for competitive advantage in firms. Organizational Behavior and Human Decision Processes 82: 150–169.

Argyris, C. 1996. Unrecognized defenses of scholars: impact on theory and research. Organization Science 7: 79–87.

Bourdieu, P. 1977. Outline of a Theory of Practice. Cambridge, U.K.: Cambridge University Press.

Brown, J.S., and P. Duguid. 1991. Organizational learning and communities-of-practice: toward a unified view of working, learning, and innovation. Organization Science 2: 40–57.

Carlile, P.R. 2002. A pragmatic view of knowledge and boundaries: boundary objects in new product development. Organization Science 13: 442–455.

Carroll, J.S. 1995. Incident reviews in high-hazard industries: sense-making and learning under ambiguity and accountability. Industrial and Environmental Crisis Quarterly 9: 175–197.

Carroll, J.S. 1998. Organizational learning activities in high-hazard industries: the logics underlying self-analysis. Journal of Management Studies 35: 699–717.

Carroll, J.S., and S. Hatakenaka. 2001. Driving organizational change in the midst of crisis. MIT Sloan Management Review 42: 70–79.

Carroll, J.S., J.W. Rudolph, and S. Hatakenaka. 2002. Learning from experience in high-hazard organizations. Research in Organizational Behavior 24: 87–137.

Carroll, J.S., J.W. Rudolph, S. Hatakenaka, T.L. Wiederhold, and M. Boldrini. 2001. Learning in the Context of Incident Investigation Team Diagnoses and Organizational Decisions at Four Nuclear Power Plants: Linking Expertise and Naturalistic Decision Making, E. Salas and G. Klein, eds. Mahwah, N.J.: Lawrence Erlbaum.

Cook, S.D.N., and J.S. Brown. 1999. Bridging epistemologies: the generative dance between organizational knowledge and organizational knowing. Organization Science 10: 381–400.

Crossan, M.M., H.W. Lane, and R.E. White. 1999. An organizational learning framework: from intuition to institution. Academy of Management Review 24: 522–537.

deGeus, A. 1988. Planning as learning. Harvard Business Review 66(2): 70–74.

Dougherty, D. 1992. Interpretive barriers to successful product innovation in large firms. Organization Science 3: 179–202.

Gruenfeld, D.H., P.V. Martorana, and E.T. Fan. 2000. What do groups learn from their worldliest members: direct and indirect influence in dynamic teams. Organizational Behavior and Human Decision Processes 82: 45–59.

Hansen, M.T. 1999. The search-transfer problem: the role of weak ties in sharing knowledge across organization subunits. Administrative Science Quarterly 44: 82–111.

Hendershot, D.C., A.G. Keiter, J.W. Kacmar, P.C. Magee, P.C. Morton, and W. Duncan. 2003. Connections: how a pipe failure resulted in resizing vessel emergency relief systems. Process Safety Progress 22(1): 48–56.

Jackson, S.A. 1996. Challenges for the Nuclear Power Industry and Its Regulators: The NRC Perspective. Speech presented at the Regulatory Information Conference, Washington, D.C., April 9, 1996.

Jackson, S.E. 1992. Team Composition in Organizational Settings: Issues in Managing an Increasingly Diverse Workforce. Pp. 138–173 in Group Process and Productivity, S. Worshel, W. Wood, and J.A. Simpson, eds. Newbury Park, Calif.: Sage Publications.

Jehn, K.A., G.B. Northcraft, and M.A. Neale. 1999. Why differences make a difference: a field study of diversity, conflict, and performance in workgroups. Administrative Science Quarterly 44: 741–763.

Kemeny, J.G., B. Babbitt, P.E. Haggerty, C. Lewis, P.A. Marks, C.B. Marrett, L. McBride, H.C. McPherson, R.W. Peterson, T.H. Pigford, T.B. Taylor, and A.D. Trunk. 1979. Report of the President's Commission on the Accident at Three Mile Island. New York: Pergamon Press.

Klein, G. 1998. Source of Power: How People Make Decisions. Cambridge, Mass.: MIT Press.

Lanir, Z. 1986. Fundamental Surprise. Eugene, Ore.: Decision Research.

March, J.G. 1991. Exploration and exploitation in organizational learning. Organization Science 2: 71–87.

Marcus, A.A., P. Bromiley, and M. Nichols. 1989. Organizational Learning in High Risk Technologies: Evidence from the Nuclear Power Industry. Discussion Paper #138. Minneapolis: University of Minnesota Strategic Management Research Center.

Marcus, A.A., and M.L. Nichols. 1999. On the edge: heeding the warnings of unusual events. Organization Science 10: 482–499.

Nutt, P.C. 1999. Surprising but true: half the decisions in organizations fail. Academy of Management Executive 13: 75–90.

Reason, J. 1990. Human Error. New York: Cambridge University Press.

Reason, J. 1997. Managing the Risks of Organizational Accidents. Brookfield, Vt: Ashgate Publishers.

Rochlin, G.I., and A. von Meier. 1994. Nuclear power operations: a cross-cultural perspective. Annual Review of Energy and the Environment 19: 153–187.

Rudolph, J.W., E.G. Foldy, and S.S. Taylor. 2001. Collaborative Off-Line Reflection: A Way to Develop Skill in Action Science and Action Inquiry. Pp. 405-412 in Handbook of Action Research, P. Reason and H. Bradbury, eds. Thousand Oaks, Calif.: Sage Publications.

Schulman, P.R. 1993. The negotiated order of organizational reliability. Administration and Society 25: 353–372.

Senge, P. 1990. The Fifth Discipline. New York: Doubleday.

Sitkin, S.B. 1992. Learning through failure: the strategy of small losses. Research in Organizational Behavior 14: 231–266.

Szulanski, G. 1996. Exploring internal stickiness: impediments to the transfer of best practices within the firm. Strategic Management Journal 17: 27–43.

van der Schaaf, T.W., D.A. Lucas, and A.R. Hale, eds. 1991. Near Miss Reporting as a Safety Tool. Oxford, U.K.: Butterworth-Heinemann.

von Hippel, E. 1994. "Sticky information" and the locus of problem solving: implications for innovation. Management Science 40(4): 429–439.

Weick, K.E., K.M. Sutcliffe, and D. Obstfeld. 1999. Organizing for high reliability: processes of collective mindfulness. Research in Organizational Behavior 21: 81–123.

Weick, K.E., and K.M. Sutcliffe. 2001. Managing the Unexpected: Assuring High Performance in an Age of Complexity. San Francisco: Jossey-Bass.

White, K.B. 1984. MIS project teams: an investigation of cognitive style implications. MIS Quarterly 8(2): 95–101.

Wildavsky, A. 1988. Searching for Safety. New Brunswick, N.J.: Transaction Press.

Section V
Linking Risk Assessment and Risk Management

Cross-Industry Applications of a Confidential Reporting Model

LINDA J. CONNELL
Human Factors and Research Technology Division
NASA Ames Research Center

A strong emphasis on public safety in the United States is apparent in many arenas of public life. The efforts toward preventing accidents is especially prominent in critical-outcome environments, where if a mistake is made, there can be tragic results. The loss of life and substantial injury that may result from accidents is especially tragic if it is discovered in the process of an investigation that the event could have been prevented. In large, complex, and dynamic environments like aviation, nuclear power, medicine, and other industries where sometimes minor errors or flaws in systems can lead to serious incidents or accidents, the challenge of maintaining safety is significant. Therefore, effective risk management, which includes risk assessment and risk mitigation, is crucial to ensuring safety.

THE AVIATION SAFETY REPORTING SYSTEM

The U.S. aviation community and the public have benefited from a historic Interagency Agreement that was signed in 1976 between the Federal Aviation Administration (FAA) and the National Aeronautics and Space Administration (NASA). This cooperative agreement was in part a response to an aircraft accident in 1974 that was the result of an ambiguous and misunderstood communication between air traffic control (ATC) and a flight crew. The flight crew descended too soon and hit a mountain in what is called a controlled-flight-into-terrain accident. In the accident investigation by the National Transportation Safety Board (NTSB), it was discovered that another airline, six weeks prior to the accident under investigation, had also misunderstood the ATC

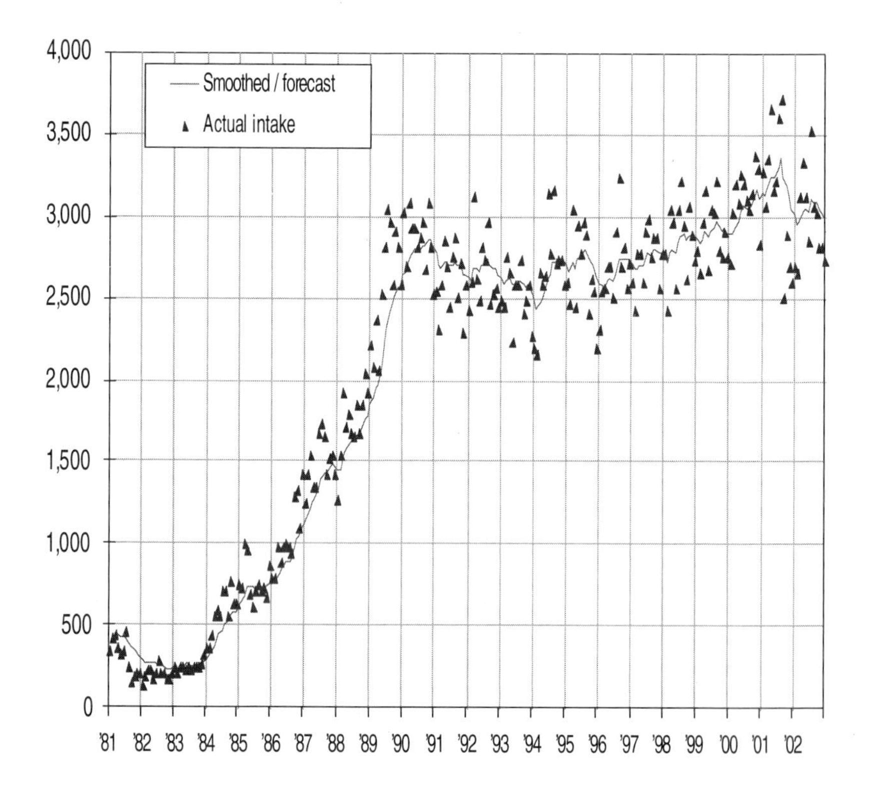

FIGURE 1 Report intake by month (1981 to 2002). Current average is 2,900 reports per month.

instruction, begun their descent, and barely missed the mountain. Although that airline had quickly warned its flight crews of the problem, other airlines were not informed. It was "an accident waiting to happen." During this investigation, the aviation industry and the government agreed that the country required a reporting system for near-misses. The FAA and NASA established the voluntary, confidential, and non-punitive reporting program entitled the Aviation Safety Reporting System (ASRS) (Reynard et al., 1986). The FAA provided immunity to aviation personnel who agreed to report to NASA under the new program (FAA, AC 00-46D).

Since that time, ASRS has accepted almost 580,000 reports from pilots, air traffic controllers, flight attendants, maintenance technicians, and others describing aviation safety events that they experienced or witnessed (Figure 1). ASRS has processed this information and contributed to the improvement of aviation safety throughout the United States and abroad (Reynard, 1991). In aviation,

ASRS has been recognized, both domestically and internationally, as a model for collecting unique safety data from frontline personnel.

Currently, seven countries besides the United States are operating aviation safety reporting systems modeled after the original ASRS, and many other countries are working to establish systems. The value of confidentiality, contributions to aviation safety, and the ability to gather information often not reported through other avenues, was quickly recognized by the United Kingdom and soon after by Canada and Australia. ASRS meets annually with these countries to coordinate and compare information concerning worldwide aviation safety through the International Confidential Aviation Safety Systems (ICASS), a group formed in 1988 that has since been recognized by the International Civil Aviation Organization (ICAO). In the ICAO Annex 13 documents, member countries throughout the world are encouraged to initiate and operate systems similar to those used by ICASS countries. New countries are referred to ICASS for assistance in designing and implementing new systems.

CROSS-INDUSTRY APPLICATIONS

The confidential reporting model has developed and matured for more than 25 years through collaboration between NASA Ames Research Center, ASRS, and the FAA Office of System Safety. The system has been recognized for providing unique safety information not available through any other means (Connell, 2000, 2002). Other disciplines and industries that have recognized the advantage of ASRS have consulted with ASRS to assess its relevance and potential contributions to their own safety efforts. The nuclear power industry has adopted a similar approach to gathering safety information to complement its traditional data-collection methods. The maritime industry is currently considering the best application of the confidential reporting model to its environment (Connell and Mellone, 1999).

Medical Reporting

The medical community has begun a strong initiative to adopt the ASRS model to collect safety information from frontline medical personnel. In *To Err Is Human: Building a Safer Healthcare System*, the Institute of Medicine (IOM) directly addresses the ASRS model (IOM, 2000).

In 1997, prior to the release of the IOM report, the Department of Veterans Affair (VA) asked NASA ASRS to join an expert advisory panel being convened in Washington, D.C., to advise the VA as they began a new focus on patient safety. The VA invited numerous cross-industry participants to describe how their industries addressed safety and which methods had been successful. At those meetings, the VA asked NASA's ASRS director if assistance could be provided to the VA to create a medical reporting system modeled after ASRS. The offer was enthusiastically accepted, and NASA entered into an interagency agreement with the VA in May 2000 to establish a collaboration between ASRS at NASA Ames Research Center at Moffett Field, California, and the VA National Center for Patient Safety (NCPS) in Ann Arbor, Michigan. The new system, the Patient Safety Reporting System (PSRS), which replicates ASRS, is the proof-of-concept for medicine; the model is expected to evolve to meet the safety needs of this complex environment (McDonald and Connell, 2001). The PSRS is in operation and is receiving reports that are providing constructive safety information. The VA and NCPS have introduced numerous safety innovations in recent years, and the PSRS is expected to be complementary to those efforts (Weeks and Bagian, 2000). The PSRS is expected to provide benefits to health-care delivery similar to ASRS's benefits to aviation.

The resources of both NASA and the VA, and strong VA protections of data from legal discovery under 38 U.S.C. 5705, have enabled the confidential reporting model to flourish and grow in medicine. The NASA Ames Research Center, the Center of Excellence in Information Technology development for NASA, has a world-renowned group of researchers in human factors. All of the technology development and human factors research that have supported ASRS are available to the NASA/VA PSRS project.

In addition, substantial developments in automated report processing, data mining, textual analysis, and data visualization tools have been made. These software and hardware tools are human-centered; that is, they support the human analysts who are essential to the success of the ASRS model. These developments and the adaptation of the aviation model to the medical environment have already contributed to patient safety and a knowledge base for proactive change.

Security Reporting

A new project is being initiated to create a separate avenue of reporting for security events. Since September 11, ASRS has received increasing numbers of reports describing aviation security incidents. But a gap analysis and study of these reports revealed that these reports were extremely sensitive and would require different methods of analysis and evaluation. In addition, although ASRS hears from pilots, air traffic controllers, flight attendants, and mechanics, other groups of personnel involved directly with the security processes have not been exposed to or educated about the confidential reporting model of ASRS. Therefore, NASA is proposing that a security incident reporting system (SIRS) be part of a new NASA program, the Aviation Safety and Security Program (AvSSP). The proposed SIRS project would be a replication of the ASRS model with all of the essential success criteria of the original model. However, because of the unique nature of this type of reporting, SIRS will probably provide alternative processing features that include more extensive protections. A consortium of industry and government stakeholders will be created to advise NASA during the development of SIRS.

RISK MANAGEMENT

The importance of risk management in high-reliability systems and industries cannot be overstated. Many concepts and methods have been proposed for effective risk management. Risk management can be defined as "the organized process of identifying and assessing risks, then establishing a comprehensive plan to prevent or minimize harmful effects from those risks being asserted" (NASA, NPG 2810.1). One method is to perform risk management during all of the life-cycle phases in the development of a new technology. In NASA guidance for research and development, risk management encompasses risk assessment, risk mitigation, evaluation of residual risk, and risk acceptance. The definition of risk used in this guidance is "a function of the probability of a given threat source exercising a particular vulnerability and the resulting impact of that adverse event on the organization" (NASA, NPG 2810.1). In high-reliability industries, where the impact of an incident can have catastrophic results, risk must be considered in relation to "threat sources" that capitalize on system "vulnerabilities."

The voluntary, confidential, non-punitive model for the reporting of safety events is a significant tool in risk management. One NASA approach to total risk management includes nine steps in the risk assessment process (Table 1).

The confidential reporting model is most useful for threat identification (Step 2) and vulnerability identification (Step 3). The stated objectives of ASRS are: (1) to identify deficiencies and discrepancies within the aviation system; and (2) to provide data and information for system planning and improvement (Connell, 2002; Reynard, 1991). In addition, ASRS is a national resource that provides three types of information: (1) identification of problems and issues in aviation systems; (2) the generation of hypotheses for further research; and (3) information about unique human and operational factors. Thus, ASRS is well situated to provide information on risk in terms of both threats and vulnerabilities.

TABLE 1 Nine Steps of Risk Assessment Model in NASA NPG 2810.1

Risk Assessment
1. System characterization
2. Threat identification
3. Vulnerability identification
4. Control analysis
5. Probability determination
6. Impact analysis
7. Risk determination
8. Control recommendations
9. Results documentation

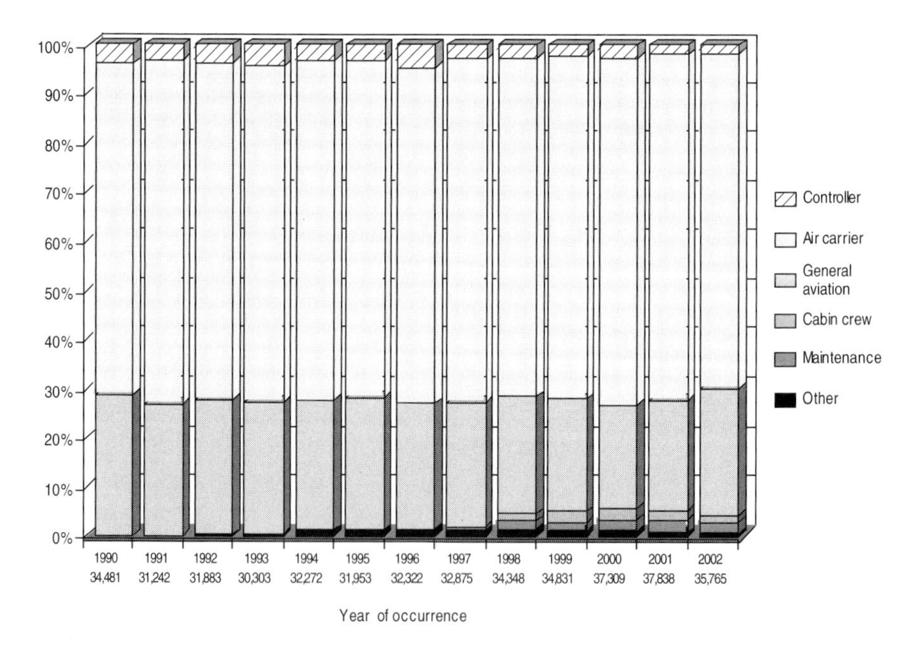

FIGURE 2 Distribution of ASRS incident reports, January 1990 to December 2002.

Because ASRS is independently administered, reporter confidentiality is protected, and the system is non-punitive, people on the front line of aviation can report in a protected environment. People who work in the system every day freely provide candid and introspective reports about their performance, whether they performed well, or not so well, in the complex aviation system. The information they volunteer describes activities and events that precede sometimes serious incidents. In reading and analyzing these reports, specialists in aviation transform the report data into information that can be used to assess risk in the system.

Because of the conditions of reporting and limited immunity established between NASA and the FAA, ASRS is a robust source of information for both threat and vulnerability identification, its main contribution to risk assessment and, thus, to risk management. By using de-identifying policies and procedures, ASRS has preserved the confidentiality of reporters for more than 25 years of successful operation. ASRS has established a reputation of trustworthiness that encourages honest, open reporting. Currently, ASRS receives approximately 38,000 reports per year (Figure 2).

Based on trust and confidence, frontline personnel have provided high-volume, high-quality, candid reports that have identified many threats and vulnerabilities. But these reports have not only revealed system weaknesses that could, combined with other factors, lead to serious incidents or accidents. They have also provided clues to some strengths in the system. ASRS reports have enabled analysts to discover how people detect anomalies in the system and how they recover from potentially dangerous events and avert fatal accidents. The people involved can then discuss thoroughly the event from the beginning. In the trusted and protected environment of ASRS, these individuals are willing to explain their roles in the occurrence. Their insights, the "human factors content," have made ASRS data valuable for improving aviation safety.

ASRS attempts to maintain a neutral, unbiased position between the numerous factions in aviation. The information generated by the system and distributed through a variety of products is provided to the government and industry aviation safety community, which develops and acts on safety solutions. ASRS often states that "it works through the good offices of others."

The contribution of ASRS information to risk management is largely through threat and vulnerability identification and descriptions of the context in which incidents occur. ASRS does not monitor or demand corrective action in the aviation system in response to the information it provides. To preserve its role as an independent, external, and neutral contributor to safety improvement, ASRS remains outside the ongoing process. Perceptions of bias, however subtle, can adversely affect people's willingness to report. The trust and the voluntary nature of ASRS are unequivocally protected.

Responses to the threats and vulnerabilities identified by ASRS and risk management are developed through mechanisms outside of ASRS, although ASRS can provide a neutral forum for continuing discussions to reduce threats and vulnerabilities.

SUMMARY

ASRS is a proven, effective system for confidential reporting and an exemplary system for application in other industries interested in safety improvements. This model, where the "devil is in the details," can be replicated, adapted, and evolved to be an intuitive, productive, information-collection mechanism for safety improvement in any system. ASRS's biggest contribution is in the identification of threats and vulnerabilities.

ASRS's characteristics and features are unique among other information-gathering systems. But its success requires constant nurturing, support, and advocacy. For people to feel that they can safely report what actually happened and happens in a system, trust and confidence can never be sacrificed to other interests. When frontline personnel in a system believe and trust that they are protected, even if they are the bearers of bad news about system flaws or they expose their own errors in the interest of system integrity, they will provide truly rich and illuminating data that can lead to safety improvements.

REFERENCES

Connell, L.J. 2000. Aviation Incident Reporting: Valuable Information for Safety. Presented at National Symposium for Building Systems That Do No Harm: Advancing Patient Safety Through Partnership and Shared Knowledge, June 29, 2000, Dallas, Texas.

Connell, L.J. 2002. Aviation Safety Reporting System: Contribution of Confidential Reporting to Aviation Safety. Presented to Institute of Medicine, Committee on Patient Safety Data Standards, September 23, 2002, Washington, D.C.

Connell, L.J., and V.J. Mellone. 1999. Aviation Safety Reporting System: A Blueprint for Maritime Safety. Presented to Society of Naval Architects and Marine Engineers, Human Factors Panel, February 4, 1999, San Francisco, California.

Federal Aviation Administration. 1997. Aviation Safety Reporting Program. Advisory Circular 00-46D. Washington, D.C.: Federal Aviation Administration.

IOM (Institute of Medicine). 2000. To Err Is Human: Building a Safer Healthcare System, L.T. Kohn, J.M. Corrigan, and M.S. Donaldson, eds. Washington, D.C.: National Academy Press.

McDonald, H., and L.J. Connell. 2001. Patient Safety Reporting System: The Who, What, Where, and Why. Presentation to 4th Annual Meeting, National Association of Inpatient Physicians, March 27, 2001, Atlanta, Georgia.

Reynard, W.D. 1991. The Acquisition and Use of Incident Data: Investigating Accidents Before They Happen. ASRS Internal Publication. Washington, D.C.: National Aeronautics and Space Administration.

Reynard, W.D., C.E. Billings, E.S. Cheaney, and R. Hardy. 1986. The Development of the NASA Aviation Safety Reporting System. NASA Reference Publication 1114. Washington, D.C.: National Aeronautics and Space Administration.

Weeks, W.B., and J.P. Bagian. 2000. Developing a culture of safety in the Veterans Health Administration. Effective Clinical Practice 3(6): 270–277.

Stuck on a Plateau
A Common Problem

CHRISTOPHER A. HART
Office of System Safety
Federal Aviation Administration

After declining significantly for about 30 years to a commendably low rate, the rate of fatal accidents for commercial aviation worldwide has been stubbornly constant for many years (Boeing Commercial Airplane Group, 2003). As part of the effort to address this problem, the Federal Aviation Administration (FAA) Office of System Safety proposed the establishment of a global aviation information network (GAIN), a voluntary, privately owned and operated network of systems that collects and uses aviation safety information about flight operations, air traffic control operations, and maintenance to improve aviation safety.[1] The proactive use of information has been greatly facilitated by technological advances that have improved the collection and use of information about adverse trends that may be precursors of future mishaps.

In the course of developing GAIN, the Office of System Safety noted that many other industries have also experienced a leveling off of mishap rates. These industries include other transportation industries, health care, national security, chemical manufacturing, public utilities, information infrastructure protection, and nuclear power. Not satisfied with the status quo, most of these industries are attempting to find ways to start their rates going down again.

Most of the industries have robust backups, redundancies, and safeguards in their systems so that most single problems, failures, actions, or inactions

[1]The FAA first published the GAIN concept in the Federal Register (61 F.R. 21522, May 10, 1996). It was then called the Global Analysis and Information Network, but the name was later changed to reflect that GAIN was created to provide tools and processes for proactive use by the aviation community, rather than to conduct analyses for the aviation community.

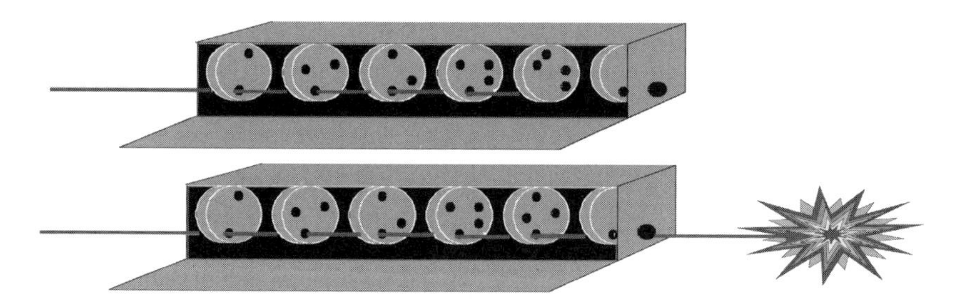

FIGURE 1 Graphic representation of potential mishaps. Source: adapted from Reason, 1990.

usually do not result in harm or damage. Several things must go wrong simultaneously, or at least serially (as "links in an accident chain"), for harm or damage to occur. That, of course, is the good news. The bad news is that the absence of a single weak point means there is no single, easily identifiable point at which to intervene.

This scenario can be represented graphically as a box containing several disks with holes spinning about a common axis (Figure 1). A light shining through the box represents a potential mishap, with each disk being a defense to the mishap. Holes in the disks represent breaches in the defenses. If all of the breaches happen to "line up," the light emerges from the box, and a mishap occurs. This figure is adapted from the Swiss cheese analogy created by James Reason (1990) of Manchester University in the United Kingdom. In Reason's analogy, a mishap occurs when the holes line up in a stack of cheese slices.

Each spinning disk (or slice of cheese) can be compared to a link in the chain of events leading to an accident. Each link, individually, may occur relatively frequently, without harmful results; but when the links happen to combine in just the wrong way—when the holes in the spinning wheels all happen to line up—a mishap occurs.

In systems with robust defenses against mishaps, the characteristic configuration of incidents and accidents can be depicted by a Heinrich pyramid (Figure 2), which shows that, for every fatal accident, there will be three to five nonfatal accidents and 10 to15 incidents; but there will also be *hundreds* of unreported occurrences (the exact ratios vary).

Usually, occurrences are not reported because, by themselves, they are innocuous (i.e., they do not result in mishaps). Nevertheless, unreported occurrences are the "building blocks" of mishaps; if they happen to combine with other building blocks from the "unreported occurrences" part of the pyramid, they may someday result in a mishap.

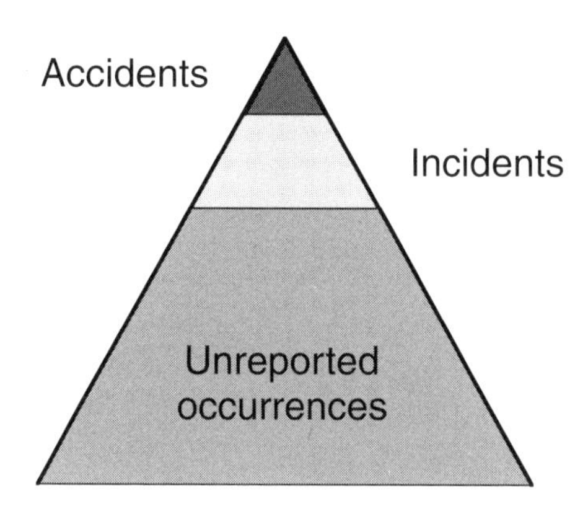

FIGURE 2 The Heinrich pyramid. Source: Heinrich, 1931.

COLLECTION AND ANALYSIS OF INFORMATION

Most industries have begun to consider the feasibility of collecting and analyzing information about precursors before they result in mishaps. Too often, the "hands-on" people on the "front lines" note, *after* a mishap, that, they "all knew about that problem." The challenge is to collect the information "we all know about" and do something about it *before* it results in a mishap.

Many industries have instituted mandatory reporting systems to collect information. They generally find, however, that there is no reasonable way to mandate the reporting of occurrences that do not rise to the level of mishaps or potential regulatory violations. Short of a mishap, the system must generally rely upon *voluntary* reporting, mostly from frontline workers, for information about problems. In the aviation industry, reporting about events near the top of the pyramid is generally mandatory, but reporting most events in the large part of the pyramid is generally voluntary. In most industries, including aviation, most of the information necessary for identifying precursors and addressing them is likely to be in the large part of the pyramid.

Legal Deterrents to Reporting

In the United States, four factors have discouraged frontline workers, whose voluntary reporting is most important, from coming forth with information. First, potential information providers may be concerned that company management and/or regulatory authorities will use the information for punitive or enforcement purposes. Thus, a worker might be reluctant to report a confusing process, fearing that management or the government might not agree that the process is confusing and might punish the worker. A second concern is that reporting potential problems to government regulatory agencies may result in the information becoming accessible to the public, including the media, which could be embarrassing, bad for business, or worse. A third major obstacle to the collection of information is the possibility of criminal prosecution. The fourth concern, perhaps the most significant factor in the United States, is that collected information may be used against the source in civil litigation.

Most or all of these legal issues must be addressed before the collection and sharing of potential safety data can begin in earnest. GAIN's experience in the aviation industry could benefit other industries in overcoming these obstacles.

Analytical Tools

Once the legal issues have been addressed, most industries face an even more significant obstacle—the lack of sophisticated analytical tools that can "separate sparse quantities of gold from large quantities of gravel" (i.e., convert large quantities of data into useful information). These tools cannot solve problems automatically, but they can generally help experienced analysts accomplish several things: (1) identify potential precursors; (2) prioritize potential precursors; (3) develop solutions; and (4) determine whether the solutions are effective. Most industries need tools for analyzing both digital data and text data.

To identify and resolve concerns, most industries will have to respond in a significantly different way than they have in the past. The ordinary response to a problem in the past was a determination of human error, followed (typically) by blaming, retraining, and/or punishing the individual who made the last mistake before the mishap occurred.

As mishap rates stabilize, they become more resistant to improvement by blaming, punishing, or retraining. The fundamental question becomes why employees who are highly trained, competent, and proud of doing the right thing make inadvertent and potentially life-threatening mistakes that can hurt people, often including themselves (e.g., airline pilots). Blaming problems on "human error" may be accurate, but it does little to prevent recurrences of the problem. Stated another way, if people trip over a step x times per thousand, how big must x be before we stop blaming people for tripping and start focusing on the step. (Should it be painted, lit up, or ramped? Should there be a warning sign?)

Most industries are coming to the conclusion that the historic focus on individuals, although still necessary, is no longer sufficient. Instead of focusing primarily on the *operator* (e.g., with more regulation, punishment, or training), it is time that we also focus on the *system* in which operations are performed. Given that human error cannot be eliminated, the challenge is how to make the system (1) less likely to result in human error and (2) more capable of withstanding human errors without catastrophic results.

Expanding our focus to include improving the system does not mean *reducing* the safety accountability of frontline employees. It does mean *increasing* the safety accountability of many other people, such as those involved in designing, building, and maintaining the system. In health care, for example, the Institute of Medicine's Committee on Quality of Heath Care in America issued a report that concluded that as many as 98,000 people die every year from medical mistakes (IOM, 2000). The report recommends that information about potentially harmful "near-miss" mistakes (i.e., mishap precursors) be systematically collected and analyzed so that analysts can learn more about their causes and develop remedies to prevent or eliminate them. The premise of this proactive approach is described as follows:

> Preventing errors means designing the health care system at all levels to make it safer. Building safety into processes of care is a much more effective way to reduce errors than blaming individuals . . . . The focus must shift from blaming individuals for past errors to . . . preventing future errors by designing safety into the system. . . . [W]hen an error occurs, blaming an individual does little to make the system safer and prevent someone else from committing the same error.

Intense public interest in improving health care systems presents major opportunities for creating processes that could be adopted by other industries. If health care and other industries joined forces, they could avoid "reinventing the wheel," to their mutual benefit. As industries focus more on improving the system, they will find they need to add two relatively unused concepts into the analytical mix: (1) system-wide intervention and (2) human factors.

System-Wide Interventions

Operational systems involve interrelated, tightly coupled components that work together to produce a successful result. Most problems have historically been fixed component by component, rather than system-wide. Problems must be addressed not only on a component level, but also on a system-wide basis.

Human Factors

Most industries depend heavily upon human operators to perform complex tasks appropriately to achieve a successful result. As more attention is focused on the system, we must learn more about developing systems and processes that take into account human factors. Many industries are studying human-factors issues to varying degrees, but most are still at an early stage on the learning curve.

SHARING INFORMATION

Although the collection and analysis of information can result in benefits even if the information is not shared, the benefits increase significantly if the information is shared—not only laterally, among competing members of an industry, but also between various components in the industry. Sharing makes the whole much greater than the sum of its parts because it allows an entire industry to benefit from the experience of every member. Thus, if a member of an industry experiences a problem and fixes it, every other member can benefit from that member's experience without having to encounter the problem. Moreover, when information is shared among members of an industry, problems that appear to a single member as an isolated instance can much more quickly be identified as part of a trend.

The potential benefits of sharing information increase the need for more sophisticated analytical tools. Since there is little need, desire, or capability for sharing raw data in most industries, data must first be converted into useful information for sharing to be meaningful. If data are not first converted into useful information, sharing will accomplish little.[2]

Thus, industry, governments, and labor must work together to encourage the establishment of more programs for collecting and analyzing information and to encourage more systematic sharing of information. Governments can facilitate the collection and sharing of information by ensuring that laws, regulations, and policies do not discourage such activities and by funding research on more effective tools for analyzing large quantities of data.

BENEFITS OF COLLECTING, ANALYZING, AND SHARING INFORMATION

More systematic collection, analysis, and sharing of information is a win-win strategy for everyone involved. Private industry wins because there will be fewer mishaps. Labor wins because, instead of taking the brunt of the blame and the punishment, frontline employees become valuable sources of information about potential problems and proposed solutions to accomplish what everyone wants—fewer mishaps. Government regulators win because the better they understand problems, the smarter they can be about proposing effective, credible remedies. Improved regulations further benefit industry and labor because effective remedies mean they will get a greater "bang for their buck" when they implement the remedies. Last but not least, the public wins because there are fewer mishaps.

[2]In most industries the shared information will also be de-identified because the benefit of sharing information about precursors usually outweighs any need to identify the source.

Immediate Economic Benefits

Whether other industries will enjoy the economic benefits that are becoming apparent in the aviation industry is still unclear. A common problem in aviation has been the difficulty encountered by airline safety professionals trying to "sell" proactive information programs to management because the cost justifications are based largely on accidents prevented. The commendably low fatal accident rate in aviation, combined with the impossibility of proving that an accident was prevented from instituted safety measures, makes it difficult, if not impossible, to prove that a company benefits financially from proactive information sharing. There *might* be an accident in the next five to seven years, and a proactive information program *might* prevent the accident that we might have. Needless to say, this is not a compelling argument for improving the "bottom line."[3]

Fortunately, an unforeseen result has begun to emerge from actual experience. The first few airlines that implemented proactive information programs quickly noticed (and reported) *immediate*, sometimes *major* savings in operations and maintenance costs as a result of information from their safety programs. It is not clear whether other industries, most notably health care, will enjoy immediate savings from information programs, but conceptually the likelihood seems high.

If immediate economic benefits can be demonstrated, this could be a significant factor in the development of mishap-prevention programs, which would become immediate, and sometimes major, profit centers, instead of merely "motherhood and apple pie" good ideas with potential, statistically predictable, future economic benefits. Moreover, the public would benefit, not only from fewer mishaps, but also from lower costs.

CONCLUSION

Many industries, including most transportation industries, health care, national security, chemical manufacturing, public utilities, information infrastructure protection, and nuclear power, have experienced a leveling off of the rate of mishaps. They are now considering adopting proactive information-gathering programs to identify mishap precursors and remedy them in an effort to lower the rate of mishaps further. It is becoming apparent that there are many common reasons for the leveling off and that many common solutions can be used by most of these industries. Although one size does not fit all, these industries have

[3]Although the insurance community may someday play a major role in efforts to be proactive, the inability to prove that accidents have been prevented may explain why, at least in the aviation industry, the insurance community has not yet been a major participant in the development of proactive information programs.

an unprecedented opportunity to work together and exchange information about problems and solutions for the benefit of everyone involved.

REFERENCES

Boeing Commercial Airplanes Group. 2003. Statistical Summary of Commercial Jet Airplane Accidents, Worldwide Operations, 1959–2002. Chicago, Ill.: The Boeing Company. Available online at *http://www.boeing.com/news/techissues/pdf/statsum.pdf.*

Heinrich, H.W. 1931. Industrial Accident Prevention. New York: McGraw-Hill.

IOM (Institute of Medicine). 2000. To Err Is Human: Building a Safer Healthcare System. Washington, D.C.: National Academy Press.

Reason, J. 1990. Human Error. Cambridge, U.K.: Cambridge University Press.

Ensuring Robust Military Operations and Combating Terrorism Using Accident Precursor Concepts

YACOV Y. HAIMES
Founding Director, Center for Risk Management of Engineering Systems
University of Virginia

Effective quantitative risk assessment and risk management must be based on systems engineering principles, including systems modeling. By exploring the commonalities between accident precursors and terrorist attacks, a unified approach can be developed to risk assessment and risk management that addresses both. In this paper, I discuss the importance of system modeling and the centrality of the state variables to vulnerability, threat, and the entire process of risk modeling, risk assessment, and risk management.

The risk assessment and risk management process is effectively based on answering two sets of triplet questions:

1. The first set represents risk assessment (Kaplan and Garrick, 1981). What can go wrong? What is the likelihood? What are the consequences?
2. The second set represents risk management (Haimes, 1991). What can be done, and what options are available? What are the trade-offs in terms of costs, benefits, and risks? What are the impacts of current policy decisions on future options?

Answering the first question in each set (what can go wrong, and what can be done, and what options are available?) requires multiperspective modeling that can identify all conceivable sources of risk and all viable risk management options. Several modeling philosophies and methods have been developed over the years that address the complexities of large-scale systems and offer various modeling schema (Haimes and Horowitz, 2003). In *Methodology for Large-Scale Systems*, Sage (1977) addresses the "need for value systems which are structurally repeatable and capable of articulation across interdisciplinary fields" for modeling the multiple dimensions of societal problems. Blauberg et al. (1977) point out that, for the understanding and analysis of a large-scale system, the fundamental principles of *wholeness* (the integrity of the system) and *hierarchy* (the internal structure of the system) must be supplemented by the principle of "the multiplicity of description for any system."

To capture the multiple dimensions and perspectives of a system, Haimes (1981) introduced hierarchical holographic modeling (HHM). Haimes argues that "To clarify and document not only the multiple components, objectives, and constraints of a system but also its welter of societal aspects (functional, temporal, geographical, economic, political, legal, environmental, sectoral, institutional, etc.) is quite impossible with a single model analysis and interpretation."

Recognizing that a system "may be subject to a multiplicity of management, control and design objectives," Zigler (1984) addressed modeling complexity in *Multifaceted Modeling and Discrete Event Simulation*, where he introduces the term *multifaceted* "to denote an approach to modeling which recognizes the existence of multiplicities of objectives and models as a fact of life." In *Synectics, the Development of Creative Capacity,* Gordon (1968) introduces an approach that uses metaphorical thinking as a means of solving complex problems. Hall (1989) develops a theoretical framework, which he calls *metasystems methodology*, with which to capture a system's multiple dimensions and perspectives.

Other seminal works in this area include *Social Systems: Planning and Complexity* by Warfield (1976) and *Systems Engineering* by Sage (1992). Sage identifies several phases in a systems-engineering life cycle to represent the multiple perspectives of a system—the structural definition, the functional definition, and the purposeful definition. Finally, in the multivolume *Systems and Control Encyclopedia: Theory, Technology, Applications,* Singh (1987) describes a wide range of theories and methodologies for modeling large-scale, complex systems.

In tracking terrorist activities as precursors to terrorist attacks, we can use HHM results to help determine which information is potentially the most worthwhile for the purposes of analysis. According to Haimes and Horowitz (2003), this information can include data related to four factors, the first three of which are areas of intelligence collection that depend on the fourth—environment:

1. the *people* associated with the potential targets (e.g., employees or members of related organizations)
2. potential *methods* of attack (e.g., specific poisons that might be most effective for a meat poisoning attack, based on detectability, potency, and ability to withstand the impact of cooking)
3. the numerous characteristics, detailed subsystems, and processes associated with potential *targets* (e.g., cybersecurity, physical security, or location)
4. *environment*, which includes information about terrorist organizations (e.g., strategies, funding, skills, cultural values, etc.) and the overall geo-

political situation (e.g., terrorist sympathizers, funding sources, training supporters, etc.).

THE STATES OF THE SYSTEM IN RISK ANALYSIS

Perrow (1999) defines an accident as "a failure in a subsystem, or the system as a whole, that damages more than one unit and in doing so disrupts the ongoing or future output of the system." We can broaden Perrow's definition of accidents to include terrorist attacks. Accidents, natural hazards, and terrorist attacks may all be perceived, or even anticipated, but still may be largely unexpected at any specific time. The causes, or initiating events, may be different for each, but the dire consequences to the system can be the same. In this paper, we use the following definitions and include both accident precursors and terrorism in each term.

State variables, the fundamental elements or building blocks of mathematical models, represent the essence of the system. State variables serve as bridges between a system's decision variables, random and exogenous variables, input, output, objective functions, and constraints. The two sets of triplet questions are bridged through state variables. Here are some examples. To control *steel production*, you must understand the state of the steel at any instant (e.g., its temperature, viscosity, and other physical and chemical properties). To know when *to irrigate and fertilize a farm* to maximize crop yield, a farmer must assess *soil moisture* and the level of *nutrients* in the soil. To *treat a patient,* a physician first must know the patient's *temperature*, *blood pressure*, and other vital states of physical health.

The word *system* connotes the natural environment, man-made infrastructures, people, organizations, hardware, and software.

In terms of terrorism, *vulnerability* connotes the manifestation of the inherent states of the system (e.g., physical, technical, organizational, cultural) that can be exploited by an adversary to cause harm or damage. However, in general terms, *vulnerability* connotes the *manifestation of the state of the system so that it could fail or be harmed by an accident-prone initiating event* (Haimes and Horowitz, 2003).

In terms of terrorism, the word *threat* connotes a potential intent to cause harm or damage to the system by adversely changing its states. However, if we generalize to include precursors to accidents, *threat* connotes *an initiating event that can cause a failure to a system or induce harm to it.* A *threat* to a vulnerable *system* results in *risk* (Haimes and Horowitz 2003).

Lowrance (1976) defines *risk* as a measure of the probability and severity of adverse effects.

If we look for the common denominator among all "accidents," including natural hazards, terrorist attacks, and human, organizational, hardware, and software failures, we find that all of them can be anticipated. However, the likelihood that they will be expected and/or realized depends on their reality and/or on our perception of (1) the state of the system being protected against accidents and (2) the associated initiating event(s).

Based on these definitions, the vulnerability of a system is directly dependent on and correlated to the levels of its state variables. Consider, for example, an electric-power infrastructure. Some of its state variables are reliability, robustness, resiliency, and redundancy (either of the entire system or of subsystems, such as generation or transmission). The task of the officials in charge is to maintain the state variables at their optimal operational level. The task of a terrorist network is to change the state variables drastically and render the system inoperable. These concepts will be discussed further in terms of the risk assessment and risk management process in the case studies that follow.

As another example, the United States is vulnerable to terrorist attacks because it has an open and free society, long borders, accessible modes of communication and transportation, and a democratic system of government. Today, these long-established state variables render it vulnerable to terrorism. A similar statement can be made about the implicit and explicit dependencies of reliability, quality, and risk on the corresponding states of the system. Therefore, a systemic, comprehensive risk assessment methodology is necessary to identify all of the system vulnerabilities; the risks associated with given threats of terrorism or accidents must also be managed through a comprehensive risk-management methodology.

The next section includes synopses of two case studies in which quantitative risk assessment and management processes were used successfully (Haimes, 2004): (1) planning, intelligence gathering, and knowledge management for military operations; and (2) the risk of a cyberattack on a water utility.

CASE STUDIES

The following synopses of two case studies demonstrate the efficacy and centrality of state variables in risk modeling, risk assessment, and risk management. In both cases HHM is used to answer the first question in the risk assessment process (What can go wrong?) and the first question in the risk management process (What can be done, and what options are available?) (Haimes, 1981, 1998). The first case study includes a brief description of HHM.

Risk Assessment and Risk Management to Support Operations Other than War

The requirements for the first line of defense against accidents in military operations include good planning, intelligence, training, and adequate resources in personnel and materiel. The following case study describes a study performed for the U.S. Army for operations other than war (OOTW) (Dombroski et al.,

2002 and CRMES, 1999). OOTW decision makers include people at all levels of the military, from strategic personnel in the Pentagon to tactical officers in the field. Recent experiences of U.S. forces involved in OOTW in Bosnia, Kosovo, Rwanda, Haiti, and now Afghanistan, Iraq, and elsewhere have dramatized the need to support military planning with information about each country that can be clearly understood. The geopolitical situation and the subject country must be carefully analyzed to support critical initial decisions, such as the nature and extent of operations and the timely marshaling of appropriate resources. Relevant details must be screened and carefully considered to minimize regrettable decisions, as well as wasted resources. Relevant details include: information on existing roads, railways, and shipping lanes; the reliability and security of electric power; communications networks; water supplies and sanitation; disease and health care; languages and cultures; police and military forces; and many others. Interagency and multinational cooperation are essential to OOTW to ensure that *ad hoc* decision making is done with careful attention to cultural, political, and societal concerns. The study developed an effective, holistic approach to decision support for OOTW that encompasses the diverse concerns affecting decision making in uncertain environments.

Hierarchical Holographic Modeling for System Characterization

There are numerous ways to characterize a country as a potential theater for OOTW. The characterization of state variables, such as technical infrastructure, political climate, society, and environment, are essential for both risk assessment and risk management. Indeed, before U.S. forces plan a deployment into a country for OOTW, the military needs to know practically *everything* important about that country. By identifying the critical state variables, as well as the state variables of U.S. forces and their allies, the military can identify (1) its own vulnerabilities (accident precursors); (2) threats from unfriendly elements; and (3) risk management options to counter these threats. HHM was the basis for the risk assessment and risk management process in the methodology developed for the Army's National Ground Intelligence Center (NGIC) and for Kosovo, as a test bed.

HHM decomposes a large-scale system into a hierarchy of subsystems, thus identifying nearly all system risks and uncertainties. Holography, a well known photographic technique, shows a multidimensional, holistic view of a subject. The holographic aspect of HHM strives to represent the many levels of the decision-making process and the inherent hierarchies in organizational, institutional, and infrastructural systems.

HHM provides a complete, documented decomposition of large-scale systems (Kaplan et al., 2001). High-level criteria in an HHM are called *head-topics*; lower level criteria relating to particular head-topics are called *subtopics.* Criteria, sources of risk, and subtopics all refer to components of the HHM. For the risk assessment process, the developers of the HHM considered the OOTW from a risk perspective, asking what could go wrong. By encapsulating nearly all possible risk scenarios, an HHM can also be perceived as the conceptualization of a "success" scenario. For the risk management process, the developers of the HHM considered the OOTW from a management perspective, asking what could be done, and what options were available.

Four HHMs were developed for OOTW: (1) the *Country HHM*, which identified a broad range of criteria for characterizing host countries and the demands they place on coalition forces; (2) the *United States (US) HHM*, which characterized what the United States had to offer countries in need; (3) the *Alliance HHM*, which characterized forces and organizations other than U.S. forces and organizations, such as multinational alliances and nongovernmental agencies; and (4) the *Objectives HHM*, which recognized the multiple, varying objectives of potential users of the methodology and coordinated the other three HHMs.

Figure 1 presents a *Country HHM* (head-topics and subtopics) based on an analysis of OOTW doctrine, case studies of previous operations, and brainstorming. Analytical case-study models from *Operation Provide Comfort*, *Operation Restore Hope*, *Operation Joint Endeavor*, and *Operation Allied Force* were analyzed to identify important criteria (C520, 1995). For example, decision makers for a typical OOTW need to know about the culture of the people, the economic and political stability of the nation, and the strength and disposition of the country's military force. For a humanitarian relief operation, they must know about the existing health-care system, as well as about the food, water, and other resources the nation can provide. For a peacekeeping mission, the focus is more on externalities and terrorists who could potentially destabilize the existing situation. In many ways, the *Country HHM* constitutes a "demand" model; it represents a country's needs in terms of personnel and materiel.

The *US HHM* addresses the supply aspect of an OOTW. The United States has a broad range of options available for addressing crisis situations, including diplomatic negotiations, economic assistance, and/or the deployment of troops and equipment. The *US HHM* is divided into two major areas: (1) *Defense Decision-Making Practice*; and (2) *Defense Infrastructure*. The *US HHM* also provides supply-side information, helping decision makers marshal supplies for an OOTW. The *Defense Infrastructure* subcriterion documents equipment, assets, and options the United States can offer to an OOTW. Details of the *US HHM* can be found in Dombroski et al. (2002).

The *Alliance HHM* is based on the recognition that the international community is more involved in maintaining international security now than it has been at any other time in world history (FM 100-8, 1997). The *Alliance HHM* documents countries, multinational alliances, and permanent and temporary relief organizations involved in an OOTW, including nongovernmental organizations (NGOs), private volunteer organizations (PVOs), and the United Nations. These organizations stabilize disengagement and ensure the economic, political, and social stability of a region after U.S. military forces leave (CALL, 1993).

Together, the *Country HMM*, *US HMM*, and *Alliance HHM* contain a vast amount of information pertaining to an OOTW, educating decision makers about the situation and helping planners and executors attain their mission goals. However, all of the information may not be important to all users at all times. A particular user may be concerned only with a subset of OOTW demands. For example, users of the system who assist in coordinating supply and demand, may be interested in a specific subset of total characterizations. The *Coordination HHM* identifies critical user-objective spaces with predictable information needs, including staff function, policy horizon, outcome valuation, and three decision-making levels: strategic, operational, and tactical. Users at each decision-making level want answers to specific questions pertaining to *Country HHM* subtopics that facilitate the identification of critical information for each decision maker.

The strategic level includes national-strategic and theater-strategic decision making. Strategic decision makers consider whether or not to enter into an operation. Operational decision makers define the operation's objectives and plan missions to maintain order and prevent escalation of the situation. Tactical decision makers plan and execute OOTW missions to support higher objectives. Details of the *Coordination HHM* can be found in Dombroski et al. (2002).

Planners must focus limited resources on the most likely sources of risk, but because of the large number of HHM risk scenarios, they may find it difficult to determine which information is important. *Risk filtering, ranking, and management* (RFRM) integrates quantitative and qualitative approaches to identify critical scenarios (Haimes et al., 2002). Using four filtering phases, decision makers can sift out the most critical 5 to 15 subtopics from the 265 subtopics in the HHM. Unfortunately, it is beyond the scope of this presentation to describe the RFRM process further.

Risk Management through Comparison Charts

The OOTW undertaken by the United States in the Balkans can illustrate the use of comparison charts, which helped determine the medical supplies needed for the incoming refugees. Officers analyzed health care and disease data for Serbia to get an understanding of the existing conditions in the province of Kosovo. To help staff officers who were not familiar with conditions in Serbia, the data were compared to data from the United States, China, and Croatia.

Figure 2 is a three-dimensional bubble chart showing three health-care metrics, one on the x axis, one on the y axis, and one in the bubble of variable areas. The figure suggests that Serbia's health-care system is in a state of disrepair—Serbia has fewer hospital physicians and beds per 1,000 people and higher infant mortality than Croatia (the United States is used as a reference base). Even though Serbia's health-care system is not as poor as China's, staff officers inferred that refugees might be in poor health, indicating that a large variety of medical supplies might be required to conduct the operation effectively.

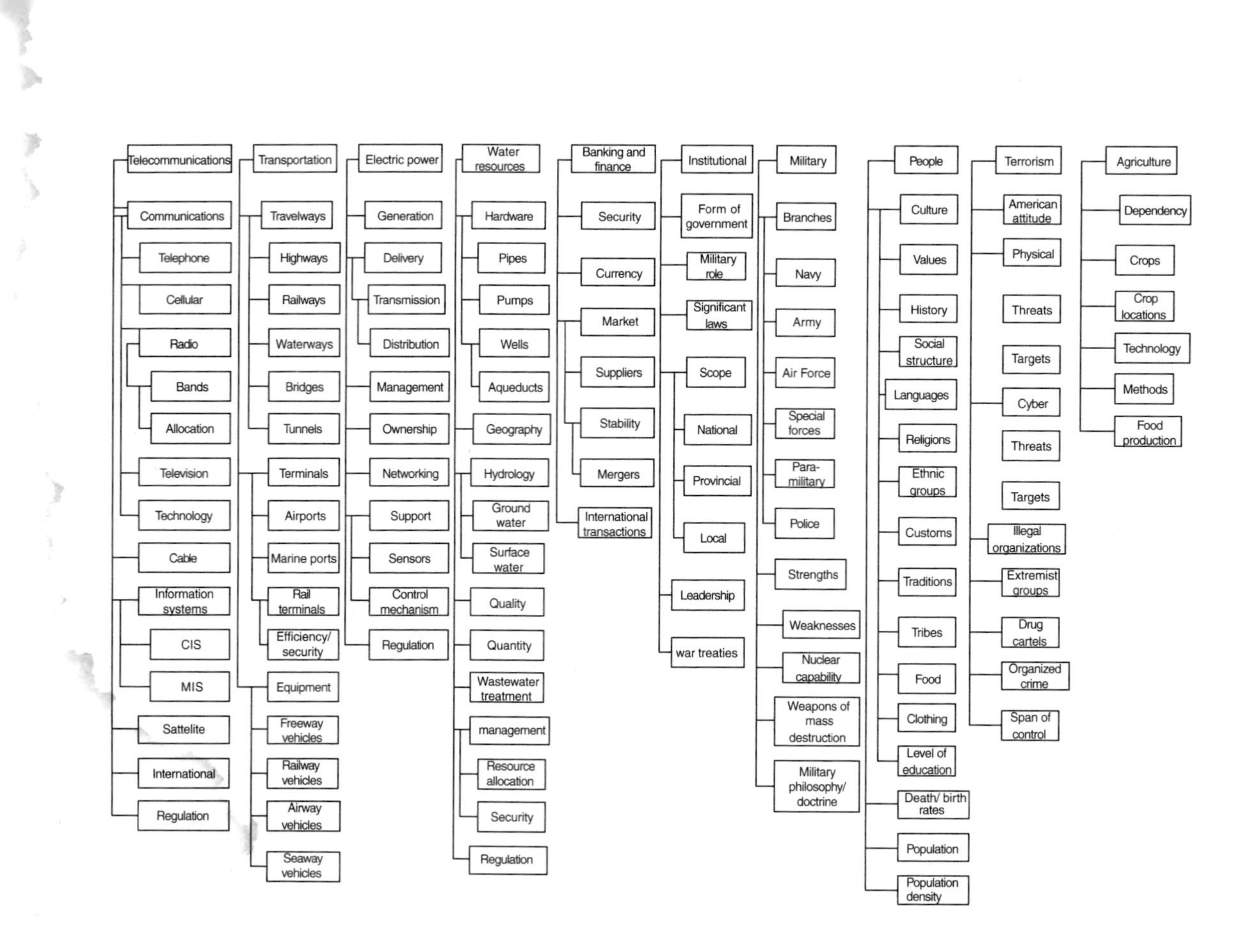
Telecommunications
Communications
Telephone
Cellular
Radio
Bands
Allocation
Television
Technology
Cable
Information systems
CIS
MIS
Sattelite
International
Regulation
Transportation
Travelways
Highways
Railways
Waterways
Bridges
Tunnels
Terminals
Airports
Marine ports
Rail terminals
Efficiency/ security
Equipment
Freeway vehicles
Railway vehicles
Airway vehicles
Seaway vehicles
Electric power
Generation
Delivery
Transmission
Distribution
Management
Ownership
Networking
Support
Sensors
Control mechanism
Regulation
Water resources
Hardware
Pipes
Pumps
Wells
Aqueducts
Geography
Hydrology
Ground water
Surface water
Quality
Quantity
Wastewater treatment
management
Resource allocation
Security
Regulation
Banking and finance
Security
Currency
Market
Suppliers
Stability
Mergers
International transactions
Institutional
Form of government
Military role
Significant laws
Scope
National
Provincial
Local
Leadership
war treaties
Military
Branches
Navy
Army
Air Force
Special forces
Para-military
Police
Strengths
Weaknesses
Nuclear capability
Weapons of mass destruction
Military philosophy/ doctrine
People
Culture
Values
History
Social structure
Languages
Religions
Ethnic groups
Customs
Traditions
Tribes
Food
Clothing
Level of education
Death/ birth rates
Population
Population density
Terrorism
American attitude
Physical
Threats
Targets
Cyber
Threats
Targets
Illegal organizations
Extremist groups
Drug cartels
Organized crime
Span of control
Agriculture
Dependency
Crops
Crop locations
Technology
Methods
Food production

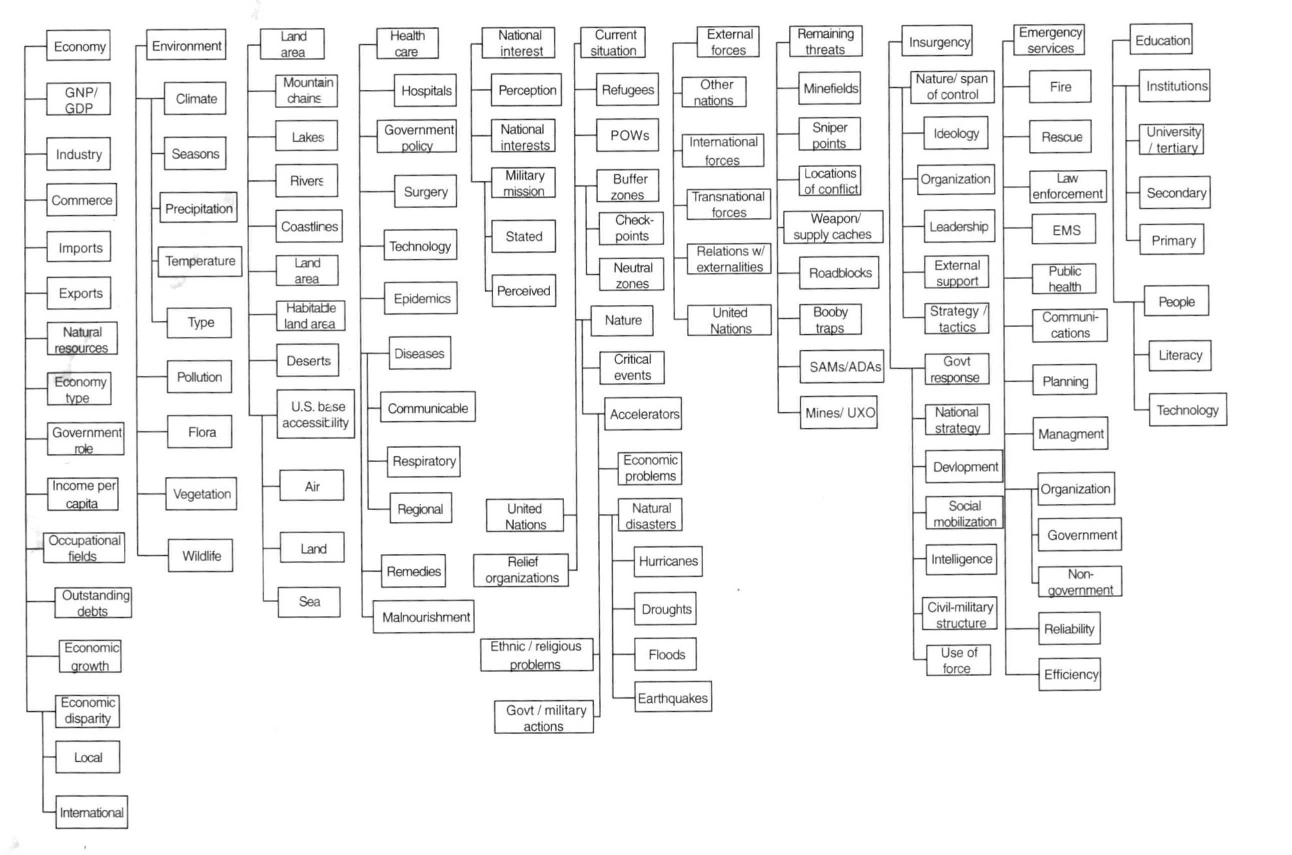

FIGURE 1 A Country HHM documenting important risks to consider about a host country for OOTW from societal, technical, political, and environmental perspectives. Source: C520, 1995.

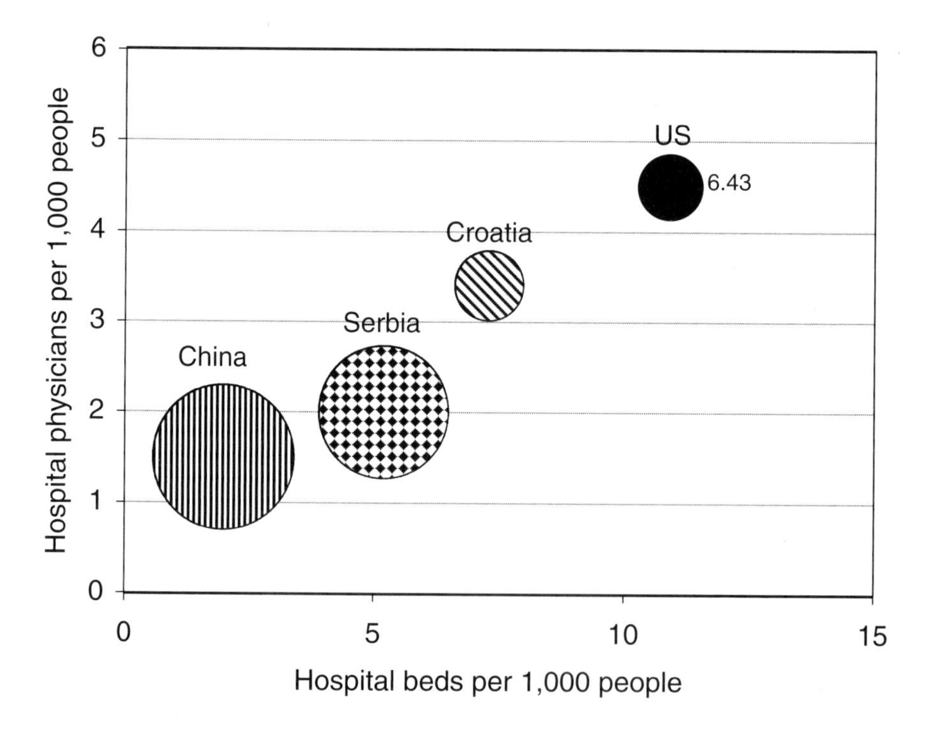

FIGURE 2 Bubble chart comparing health-care metrics for Serbia, Croatia, China, and the United States. Source: CRMES, 1999.

To determine which diseases they might encounter, the staff officers studied Figure 3, which shows the estimated prevalence of certain diseases in Serbia and Croatia. The metrics on the radials of Figure 3 indicate the percentage of the population infected. The comparison shows that the prevalence of AIDS, hepatitis A and E, and typhoid fever is higher in Serbia than in Croatia.

Risk of a Cyberattack on a Water Utility Supervisory Control and Data Acquisition Systems

Water systems are increasingly monitored, controlled, and operated remotely through supervisory control and data acquisition (SCADA) systems. The vulnerability of the telecommunications system opens the SCADA system to intrusion by terrorist networks or other threats. This case study addresses the risks of willful threats to water utility SCADA systems (Ezell, 1998; Ezell et al., 2001; Haimes, 2004). As a surrogate for a terrorist network, the focus is on a disgruntled

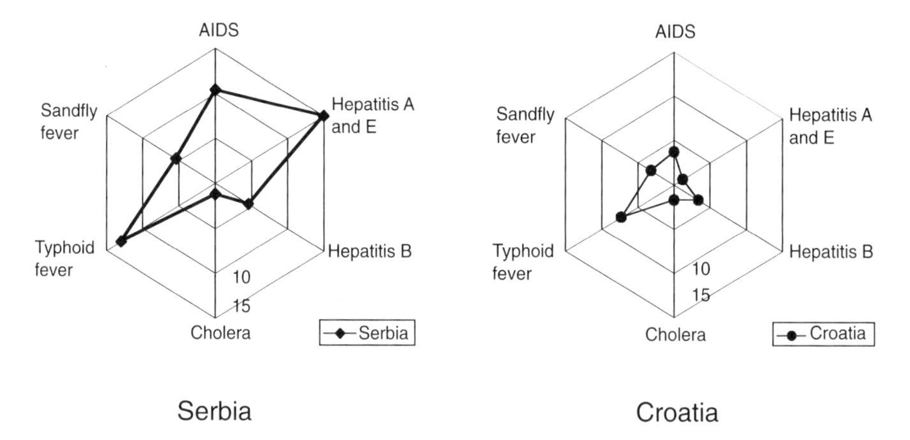

FIGURE 3 Radial chart showing the prevalence of diseases in Serbia and Croatia (numbers show the percentage of the population infected). Source: CRMES, 1999.

employee's attempt to reduce or eliminate the water flow in a city we'll call XYZ. The data are based on actual survey results showing that the primary concern of U.S. water-utility managers in XYZ was disgruntled employees.

Identifying Risks through System Decomposition

HHM yielded the following major head-topics and subtopics:

Head-Topic A: *Function*. Because of the importance of the water distribution system, its normal functioning is at major risk from cyberintrusion. This head-topic may be partitioned into three subtopics: A_1 gathering, A_2 transmitting, and A_3 distributing.

Head-Topic B: *Hardware*. SCADA hardware is vulnerable to tampering in a variety of configurations. There are nine subtopics: B_1 master terminal unit (MTU), B_2 remote terminal unit (RTU), B_3 modems, B_4 telephone lines, B_5 radio, B_6 integrated services digital network (ISDN), B_7 satellite, B_8 alarms, and B_9 sensors. Depending on the tools and skills of an attacker, hardware elements could have a significant impact on water flow for a community.

Head-Topic C: *Software*. This head-topic, perhaps the most complex, also represents the most dynamic aspects of changes in water utilities. Software has many components that are sources of risk—among them are C_1 controlling and C_2 communication.

Head-Topic D: *Human*. There are two major subtopics: D_1 employees and D_2 attackers. This head-topic addresses a decomposition of humans capable of tampering with a system.

Head-Topic E: *Tools*. A distinction is made between the various types of tools an intruder may use to tamper with a system. There are six subtopics: E_1 user command, E_2 script or program, E_3 autonomous agent, E_4 tool kit, E_5 data trap, and E_6 high-energy radio frequency (HERF) weapon (Howard, 1997).

Head-Topic F: *Access*. There are many paths (or vulnerabilities) into a system that an intruder might exploit. There are five subtopics: F_1 implementation vulnerability, F_2 design vulnerability, F_3 configuration vulnerability, F_4 unauthorized use, and F_5 unauthorized access. Even though a system may be designed to be safe, its installation and use may lead to multiple sources of risk.

Head-Topic G: *Geography*. Physical location is not a relevant factor in cyber-intrusion because tampering with a SCADA system can have global sources. International borders are virtually nonexistent because of the Internet. Four subtopics are identified: G_1 international, G_2 national, G_3 local, and G_4 internal.

Head-Topic H: *Time*. The temporal category shows how present or future decisions affect the system. For example, a decision to replace a legacy SCADA system in 10 years may have to be made today. Therefore, this head-topic addresses the life cycle of the system. There are four subtopics: H_1 long term: ≥ 10 years, H_2 short term: ≥ 5 years and < 10 years, H_3 near term: < 5 years, and H_4 today.

Measuring Risk Using the Partitioned, Multiobjective Risk Method

The partitioned, multiobjective risk method (PMRM) was used to quantify and reduce the risks of extreme events (i.e., events with low probability and high consequences) (Asbeck and Haimes, 1984; Haimes, 1998). PMRM generates a conditional expectation, given that the random variable is limited to either a range of probabilities or a range of consequences. Because the cumulative distribution function (CDF) is monotonic, there is a one-to-one relationship between the partitioned probability axis and the corresponding consequences. In this case study, we are interested in the common-unconditional expected value of risk, denoted by f_5 and in the conditional expected value of risk of extreme events (low probability/high consequences) denoted by f_4. The function f_1 denotes the cost associated with risk management.

Let the random variable Q denote water flow and q denote a realization of the random variable Q. Let $f(q)$ denote the probability density function (PDF) and β the partitioning point of water flow. Then, the conditional expected value of risk of water flow below β is:

$$f_4(\beta) = E[Q|q \leq \beta] = \frac{\int_{\beta}^{+\infty} qf(q)dq}{\int_{\beta}^{+\infty} f(q)dq}$$

Thus, for a particular policy option, there is a measure of risk f_4, in addition to the common (unconditional) expected value $E[Q]$:

$$f_5 = \int_{0}^{+\infty} qf(q)dq = E[Q]$$

City XYZ

City XYZ, a relatively small city with a population of 10,000 has a water distribution system that accepts processed and treated water "as is" from an adjacent city (Wiese et al., 1997). The water utility of XYZ is primarily responsible for an uninterrupted flow of water to its customers. The SCADA system uses a master-slave relationship, relying on the total control of the SCADA master; the remote terminal units are dumb. There are two tanks and two pumping stations as shown in Figure 4. The first tank serves the majority of customers; the second tank serves fewer customers in a topographically higher area. Tank II is located at a higher topographical point than the highest customer served. The function of the tanks is to provide a buffer and to allow the pumps to be sized lower than peak instantaneous demand.

The tank capacity has two component segments: (1) reserve storage that allows the tank to operate over a peak week when demand exceeds pumping capacity; and (2) control storage, the portion of the tank between the pump cut-out and cut-in levels. Visually, the control storage is the top portion of the tank. If demand is lower than the pump rate (low-demand periods), the level rises until it reaches the pump cut-out level. When the water falls to the tank cut-in level, it triggers the pump to start operating. If the demand is higher than the pump rate, the level continues to fall until it reaches reserve storage. The water level then remains in this area until the demand has fallen long enough to allow it to recover. The reserve storage is sized according to demand (e.g., Tank I with its larger reserve storage serves more customers).

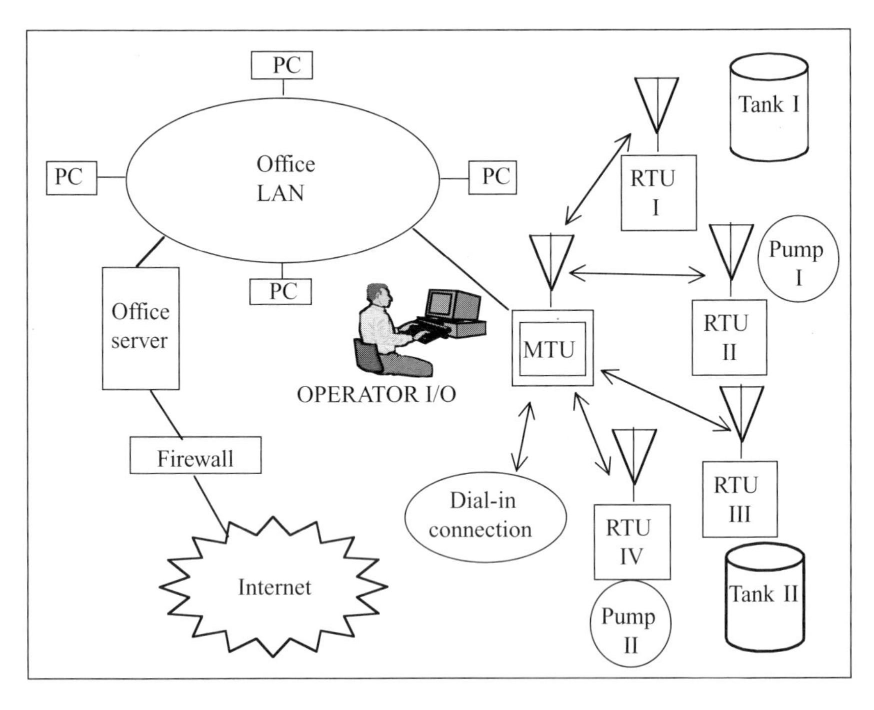

FIGURE 4 Interconnectedness of the SCADA system, local area network, and the Internet. Source: adapted from Ezell, 1998.

The SCADA master communicates directly with pumping stations I and II and signals the units when to start and stop; operating levels are kept in the SCADA master. Pumping Station I boosts the flow of water beyond the rate that can be supplied by gravity. The function of Pumping Station II is to pump water during off-peak hours from Tank I to Tank II. The primary operational goal of both stations is to maximize gravity flow and, as necessary, pump during off-peak hours as much as possible. The pumping stations receive a start command from the SCADA master via the master terminal unit (MTU) and attempt to start the duty pump. Each tank has separate inlets (from the source) and outlets (to the customer); water level and flows in and out are measured at each inlet and outlet. An altitude control valve closes the inlet when the tank is full; the "full" position is above the pump cut-out level so there is no danger of closing the valve during pumping. If something goes wrong and the pump does not shut off, the altitude valve closes, and the pump stops delivery on overpressure to prevent the main from bursting.

The SCADA system is always dependent on the communications network of the MTU and the SCADA master, which regularly polls all remote sites.

Remote terminal units (RTUs) respond only when polled to ensure that there is no contention on the communications network. The system operates automatically; the decision to start and stop pumps is made by the SCADA master and not by an operator sitting at the terminal. The system has the capability of contacting operations staff after hours via a paging system in the event of an alarm.

In the example, the staff has dial-in access, so, if they are contacted, they can dial in from home and diagnose the extent of the problem. The dial-in system has a dedicated PC connected to the Internet and the office's local area network (LAN). A packet-filter firewall protects the LAN and the SCADA. The SCADA master commands and controls the entire system, and the communications protocols for SCADA communications are proprietary. The LAN, the connection to the Internet, and the dial-in connection all use transmission-control protocol and Internet protocol (TCP/IP); instructions to the SCADA system are also encapsulated with TCP/IP. Once instructions are received by the LAN, the SCADA master de-encapsulates TCP/IP, leaving the proprietary terminal emulation protocols for the SCADA system. The central facility is organized into different access levels for the system, and an operator or technician has a level of access, depending on need.

Identifying Risks through System Decomposition

The HHM head-topics and subtopics listed earlier identify 60 sources of risk to the SCADA system of XYZ. The access points for the system are the dial-in connection points and the firewall connecting the utility to the Internet. In this example, the intruder uses the dial-in connection to gain access and control of the system.

The intruder's most likely course of action is to use a password to access the system and its control devices (Ezell, 1998). Because physical damage to equipment from dial-in access inherently requires analog fail-safe devices, managers conclude that the intruder's probable goal is to manipulate the system to affect adversely the flow of water to the city. For example, if water hammers are created, they might burst mains and damage customers' pipes. Or, the intruder might shut off valves and pumps to reduce water flow. After discussing the potential threats, the managers conclude that their greatest concern is that a disgruntled employee might tamper with the SCADA system in these ways.

Risk Management Using Partitioned, Multiobjective Risk Method

For each alternative, the managers would benefit from knowing both the expected percentage of water-flow reduction and the conditional expected extreme percentage reduction in 1-in-100 outcomes (corresponding to β). The PMRM partitions the framework s_1, s_2, s_3, ..., sn on the consequence (damage) axis at β for all alternative risk management policies. For this presentation, we used the assessment of Expert A for the event tree in Figure 5 (A. Nelson,[1] personal communication, 1998). The figure represents the state of performance of the current system based on an expert's assessment of an intruder's ability to transition through the mitigating events of the event tree. The initiating event, cyberintrusion, engenders each subsequent event, culminating with the consequences at the end of each path through the event tree. Assuming a uniform distribution of damage for each path through the tree, a composite, or mixed, probability density is generated. The uniform distribution is appropriate because the managers are indifferent beyond the upper and lower bounds.

The conditional expected value of water-flow reduction for the current system at the partitioning of the worst-case probability axis at 1-in-100 corresponds to β = 98.7 percent. Thus, the conditional expected value for this new region is 99.5 percent. Using Equations 1 and 2, five expected values of risk $E(x)$ and five conditional expected values of risk, $f_4(\beta)$, were generated and plotted (Figure 6).

Assessing Risk Using Multiobjective Trade-off Analysis

Figure 6 shows the plot of the cost of each alternative on the vertical axis and the consequences on the horizontal axis. In the unconditional expected value of risk region, Alternatives 5 and 6 are efficient. For example, Alternative 5 outperforms Alternative 3 and costs $56,600 less. In the conditional expected value of risk (worst-case region), only Alternatives 5 and 6 are efficient (Pareto optimal policies). Alternative 5 reduces the expected value of water-flow reduction by 57 percent for the 1-in-100 worst case. Note, for example, that although Alternative 5 yields a relatively low expected value of risk, at the partitioning β, the conditional expected value of risk is markedly higher (more than 40 percent). To supplement the information from our analysis, the managers use their judgment to arrive at an acceptable risk management policy.

Conclusions

This case study illustrates how risk assessment and risk management can be used to help decision makers determine preferred solutions to cyberintrusions. The approach was applied to a small city using information learned from an expert. The limitations of this approach are: (1) it relies on expert opinion to

[1]Mr. Nelson works with a wide range of applications, from business and government accounting to technical applications, such as electronic and mechanical computer-assisted drafting and has worked on a variety of standard and proprietary platforms (i.e., UNIX, PC, DOS, and networks). He implements security measures at the computer-hardware level, the operating-systems level, and the applications level.

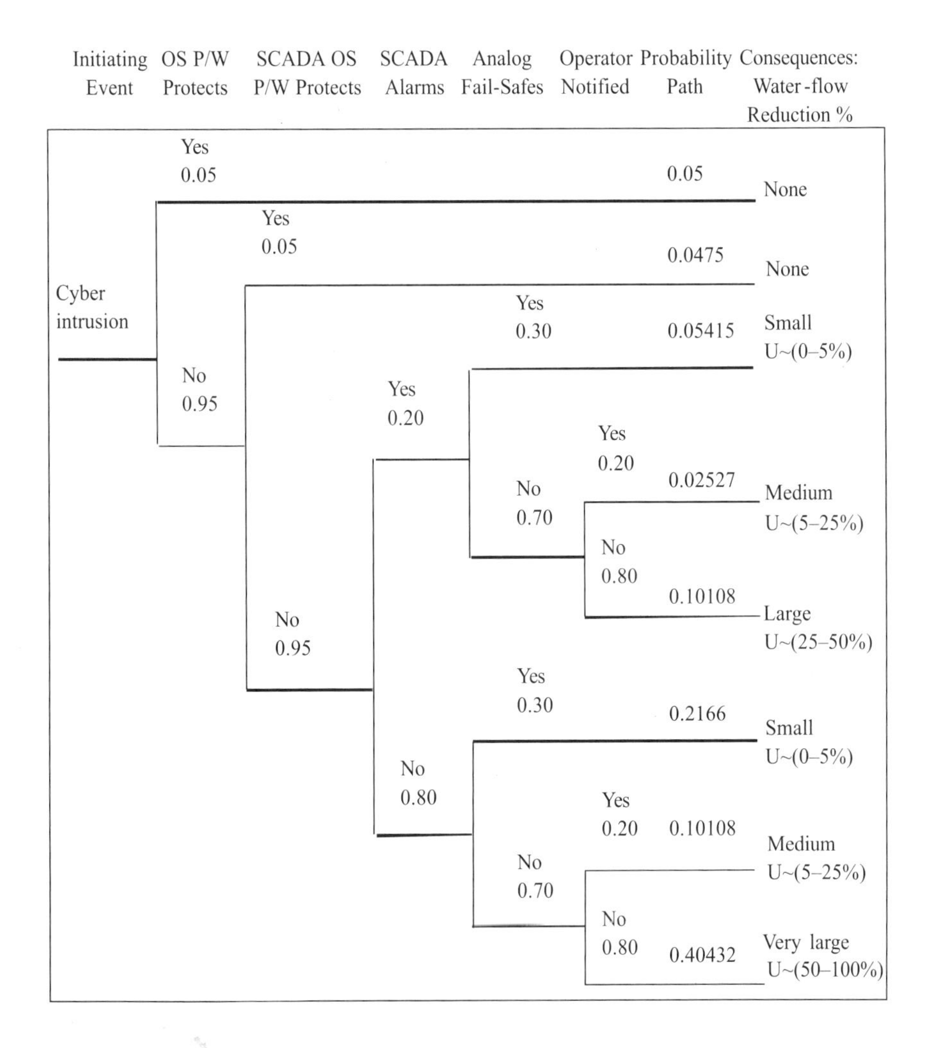

FIGURE 5 Event-tree model of the mitigating events in place to protect the system. Source: A. Nelson, personal communication.

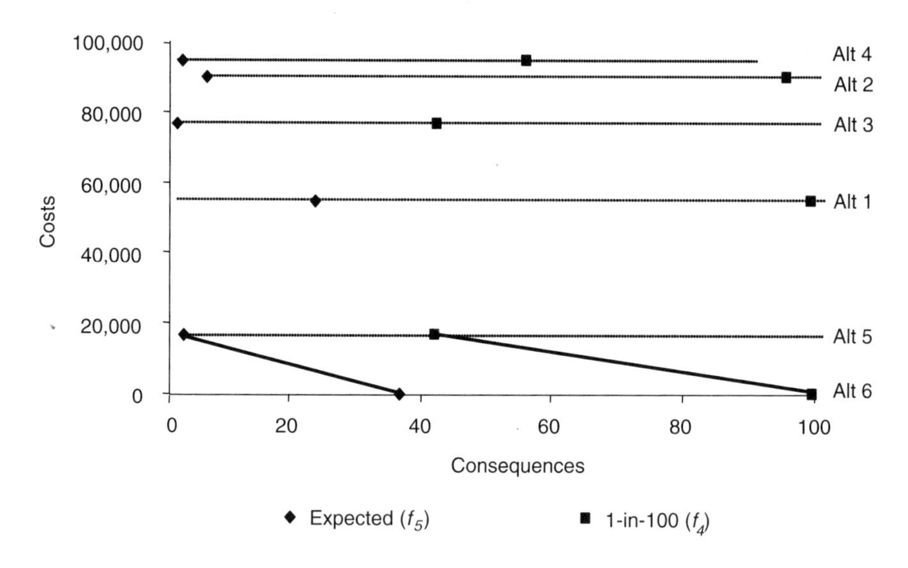

FIGURE 6 Trade-off costs for f_5 vs. f_4.

estimate probabilities for the event tree; (2) the model is not dynamic, so it does not completely represent changes in the system during a cyberattack; and (3) the event tree produces a probability mass function that must be converted to a density function for the exceedance probability to be partitioned. The underlying assumption that damages are uniformly distributed must be further explored.

REFERENCES

Asbeck E., and Y.Y. Haimes. 1984. The partitioned multi-objective risk method. Large Scale Systems 6(1): 13–38.

Blauberg, I.V., V.N. Sadovsky, and E.G. Yudin. 1977. Systems Theory: Philosophical and Methodological Problems. New York: Progress Publishers.

C520. 1995. Operations Other Than War. January 2, 1995. Fort Leavenworth, Kansas: U.S. Army Command and General Staff College.

CALL (Center for Army Lessons Learned). 1993. Operations Other Than War. Volume IV: Peace Operations. Newsletter No. 93-8. Fort Leavenworth, Kansas: U.S. Army Combined Arms Command.

CRMES (Center for Risk Management of Engineering Systems). 1999. Application of Hierarchical Holographic Modeling (HHM): Characterization of Support for Operations Other Than War. Charlottesville, Va.: Center for Risk Management of Engineering Systems, University of Virginia.

Dombroski, M., Y.Y. Haimes, J.H. Lambert, K. Schlussel, and M. Sulkoski. 2002. Risk-based methodology for support of operations other than war. Military Operations Research 7(1): 19–38.

Ezell, B. 1998. Risks of Cyber Attack to Supervisory Control and Data Acquisition for Water Supply. Masters thesis, Department of Systems Engineering, University of Virginia, Charlottesville.

Ezell, B., Y.Y. Haimes, and J.H. Lambert. 2001. Risk of cyber attack to water utility supervisory control and data acquisition systems. Military Operations Research 6(2): 23–33.

FM (Field Manual) 100-8. 1997. The Army in multinational operations. Washington, D.C.: Headquarters, Department of the Army.

Gordon, W.J.J. 1968. Synectics: The Development of Creative Capacity. New York: Collier Books.

Haimes, Y.Y. 1981. Hierarchical holographic modeling. IEEE Transactions on Systems, Man, and Cybernetics 11(9): 606–617.

Haimes, Y.Y. 1991. Total risk management. Risk Analysis 11(2): 169–171.

Haimes, Y.Y. 1998. Risk Modeling, Assessment, and Management. New York: John Wiley and Sons.

Haimes, Y.Y. 2004. Risk Modeling, Assessment, and Management, 2nd ed. New York: John Wiley and Sons.

Haimes, Y.Y., S. Kaplan, and J.H. Lambert. 2002. Risk Filtering, Ranking and Management Framework Using Hierarchical Holographic Modeling. Risk Analysis 22(2): 383–397.

Haimes, Y., and B. Horowitz. 2003. Adaptive two-player hierarchical holographic modeling game for counterterrorism intelligence analysis. Center for Risk Management of Engineering Systems. University of Virginia. Report 2003-12.

Hall, A.D., III. 1989. Metasystems Methodology: A New Synthesis and Unification. New York: Pergamon Press.

Horowitz, B., and Y.Y. Haimes. 2003. Risk-based methodology for scenario tracking for terrorism: a possible new approach for intelligence collection and analysis. Systems Engineering 6(3): 152-169.

Howard, J. 1997. An Analysis of Security Incidents on the Internet. Ph.D. dissertation, Carnegie Mellon University. Available online at *http://www.cert.org/research/JHThesis/*.

Kaplan, S., and B.J. Garrick. 1981. On the quantitative definition of risk. Risk Analysis 1(1): 11–27.

Kaplan, S., Y.Y. Haimes, and B.J. Garrick. 2001. Fitting hierarchical holographic modeling into the theory of scenario structuring and a resulting refinement to the quantitative definition of risk. Risk Analysis 21(5): 807–819.

Lowrance, W. 1976. Of Acceptable Risk. Los Altos, Calif.: William Kaufman.

Perrow, C. 1999. Normal Accidents. Princeton, N.J.: Princeton University Press.

Sage, A.P. 1977. Methodology for Large-Scale Systems. New York: McGraw-Hill.

Sage, A.P. 1992. Systems Engineering. New York: John Wiley and Sons.

Singh, M.G., ed. 1987. Systems and Control Encyclopedia: Theory, Technology, Applications. New York: Pergamon Press.

Warfield, J.N. 1976. Social Systems: Planning and Complexity. New York: John Wiley and Sons.

Wiese, I., C. Hillebrand, and B. Ezell. 1997. Scenarios One and Two: Source to No 1 PS to No 1 Tank to No 2 PS to No 2 Tank (High-Level) for a Master-Slave SCADA System. SCADA Consultants, SCADA Mail List available online at members.iinet.net.au/nianw/mailst.html.

Zigler, B.P. 1984. Multifaceted Modeling and Discrete Simulation. New York: Academic Press.

APPENDIXES

Appendix A
Letters to the Committee

Notes toward a Theory of Accident Precursors and Catastrophic System Failure

ROBERT A. FROSCH
John F. Kennedy School of Government
Harvard University

For want of a nail the shoe is lost,
for want of a shoe the horse is lost,
for want of a horse the rider is lost,
for want of a rider the battle is lost,
for want of a battle the kingdom is lost,
all for the loss of a horseshoe nail.

Benjamin Franklin, Poor Richard's Almanac *(based on a rural saying collected by English poet, George Herbert, and published in 1640)*

So, naturalists observe, a flea
Hath smaller fleas that on him prey;
And these have smaller still to bite 'em;
And so proceed *ad infinitum.*

Jonathan Swift. A Rhapsody (1733)

Given enough layers of management, catastrophe need not be left to chance.

Norman Augustine, Augustine's Laws (as recollected by this author)

In these notes, I make some observations about accidents and system failures drawn from parts of the literature that are commonly consulted for this subject. Although I do not formulate a complete and connected theory, my intent is to point out the directions from which a theory may be developed. I place this possible theory in the context of complexity and the statistical mechanics of physical phase change.

Machines and organizations are designed to be fractal. Machines are made of parts, which are assembled into components, which are assembled into subassemblies, which are assembled into subsystems, and so on, until finally they are assembled into a machine. Hierarchical organizations are specifically designed to be fractal, as any organization diagram will show. Neither a hierarchical organization nor a machine is, of course, a regular, mathematically precise fractal; they may be described as "heterogeneous fractals." Nevertheless, we expect the distribution of the number of parts of a machine vs. the masses (or volumes) of those parts to follow an inverse power law, where the power describes the dimension of the fractal (Mandelbrot, 1982). For machines, I would expect the dimension to be between two and three. For organizations, I would expect the dimension to be between one and two.

Because they are fractal, both machines and organizations are approximately (heterogeneously) scale free, that is, they look the same at any scale. (Scale free is used in the sense that: $f(kx)/f(x)$ is not a function of x. Most functions are not scale free. Power laws are scale-free functions since: $(kx)^a/x^a = k^a$.) In the case of the machine, a shrinking engineer will be surrounded by machinery at any scale; in the case of an organization, all levels of local organization are similar. (One boss has n assistant bosses, each of whom has n assistant-assistant bosses, and so on to [$(assistant)^m$] bosses.)

Many natural (and human) systems appear to develop to a self-organized critical (SOC) state (e.g., Barabosi, 2003), in which they have a scale-free fractal structure and are on the edge of catastrophe (Bak, 1996; Buchanan, 2001). Such systems appear to undergo disasters of all scales, from the miniscule to the completely destructive. The distribution of the structure of these systems is fractal, and the distribution of the (size vs. number of occurrences of a given size) catastrophe follows an inverse power law in the vicinity of catastrophe. Examples include: sandpiles (more correctly, rice piles), earthquakes, pulsar glitches, turbidite layers (fossil undersea avalanches), solar flares, sounds from the volcano Stromboli, fossil genera life spans, traffic jams, variations in cotton futures, people killed in deadly conflicts, and research paper citations. This is also the case for the ranking of words in the English language by frequency and the ranking of cities by size (Zipf, 1949).

For reasons of economy and efficiency, engineered systems (which I will loosely refer to as machines) and organizations (including those in which the design and operations of the machines are embedded) are designed to be as close to catastrophe as the designer dares. In the case of machines, the "distance" from envisioned catastrophes, called the factor of safety, varies depending upon the stresses predicted to be placed on the machine during its operating life. Organizations (as operating machines) are designed to be as lean (and mean and cheap) as seems consistent with performing their functions in the face of their operating environments. In this sense, these fractal systems may be described as design organized critical (DOC). I argue that the physics that applies to phase changes in natural SOC systems may also be applied to DOC systems.

I now introduce percolation theory, which embodies the use of the renormalization group (mean field theory) and has been used as a theoretical framework for natural phase change (Grimmett, 1999; Stauffer and Aharony, 1994). I assert that percolation theory provides a suitable "spherical cow" or "toy model" of disaster in machines and organizations. There are a number of possible percolation models, such as lattices of any dimension. I will use percolation on a Bethe lattice (Cayley Tree), although percolation on other lattices gives similar results. A Bethe lattice is a multifurcation diagram. (The simplest nontrivial case, in which multi = 2, consists of a tree of repeated bifurcations at the end of each branch.) The asymptotically, infinitely large case can be solved exactly. It has been shown (both by approximation and computation) that in less than infinite cases the phenomena, particularly around the critical value (see below), approximate the proven phenomena for the infinite case.

In our model, a link between two nodes may be conducting or nonconducting. If conducting, we regard it as a failure of that link. Strings of connected link failures are interpreted as accidents of various sizes, and a string of link failures to the central (or origin) node is interpreted as a complete catastrophe. We examine the probability of catastrophe and the distribution of lesser failures as p (the probability of failure of any link) increases from zero (Grimmett, 1999; Stouffer and Aharony, 1994).

We first would like to know the percolation threshold: the value of $p = p_c$, for which there is at least one infinite path through the infinite lattice. This may be shown to be $p_c = 1/(z-1)$ where z is multi, the number of links at each node.

Next, we would like to know P, the probability that the origin (or any other arbitrarily selected site) belongs to the infinite (or catastrophic) cluster. Stauffer and Aharony (1994) prove the example for $z = 3$:

$$p_c = \tfrac{1}{2} \qquad P = 0, \text{ for } p < p_c \qquad P/p = 1 - [(1-p)/p]^3, \text{ for } p > p_c$$

Further, it may be shown that $n_s(p)$, the average number of clusters (per site) containing s sites goes asymptotically to: $n_s(p_c) \sim s^{-\tau}$. More generally, the mean cluster size (S) near the critical threshold $(p = p_c)$ goes as $S \sim / p - p_c /^{-\gamma}$, where g is a constant. This distribution of cluster sizes describes cluster distributions near phase change in many physical systems, including the Ising model of magnetization, and clusters of water molecules near the phase change from steam to liquid water.

Catastrophic system failures are what they seem to be, phase changes, for example, from organized shuttle to rubble (a "liquid"). I note also that the percolation model leads to the suggestion that the Heinrich diagram, or occurrence pyramid is likely to be correct (Corcoran [p. 79] and Hart [p. 147] in this volume).

It is also interesting to note that Reason's "Swiss cheese model" of accidents is a percolation model, although he does not call it that or formally develop its statistical implications (Reason, 1997). Any occurrence may be an accident precursor.

What can one do? Clearly, as Hendershot has suggested in another paper in these proceedings, (p. 103 in this volume) we may redesign the system to be simpler and to function without subsystems or components that are likely to fail in a way that leads to accidents. If this is not feasible, the system must be strengthened, that is, moved farther away from its breaking point. For machines, this can be done by strengthening the elements that appear most likely to fail, and whose failure is likely to lead to disaster, by introducing redundancy, more (or larger [which is equal to more]) mechanical strength elements, redundant sensors, controls and actuators, etc. The trick is to find the elements most likely to fail, singly or in combination, so that only they are strengthened, and no "unnecessary" redundancies are introduced. The critical elements are found by engineering intuition, engineering analysis, and/or some kind of probabilistic risk assessment analysis.

In organizations, one can add people or redundant organizational elements intended to increase strength against mistakes of various kinds. These may include safety organizations, inspectorates, or auditors. However, other organizational means may also provide the necessary redundancy. In an organization with a reasonable atmosphere of trust among its members and echelons, juniors formally or informally bring their problems and troubles to their peers and their seniors, who may have other and/or broader means of attacking them. Seniors are attuned and attentive to rumors and concerns of both peers and juniors. These are strengthening elements that bridge over portions of the organizational tree and strengthen the organization. They may be likened to bringing in reinforcements. In this theoretical framework, these horizontal and vertical means of communication are strengthening elements that move the organizational structure away from p_c without adding suborganizations or people. In the scientific and engineering communities, peer review plays this communication role. The prime purpose of peer review is not to provide confirmation of excellence but to find errors and omissions that might be damaging or catastrophic.

In their work on high-reliability organizations, LaPorte, Roberts, and Rochlin (cited in Reason, 1997) describe the field reorganization of hierarchies into small teams whose members communicate directly with each other, particularly when warning of danger. For example, usually highly hierarchical Navy crews, when working together as flight deck teams on an aircraft carrier during flight operations, become a flat, highly communicating group, in which authority comes from knowledge and the perception of problems rather than from organizational position.

In summary, the statistical properties of designed machines and organizations are similar to those of natural SOC systems, and we should expect the same theoretical framework that applies to them, and to statistically similar physical phase changes, to apply to machines and organizations. Therefore, we can expect to predict the general statistical properties of accident precursors and catastrophic system failure in human-made systems from well known theoretical structures. The results also suggest why commonly used means to strengthen systems work to move the system state away from the critical p_c.

Further development and application of this theory will require applying it specifically to machine and organizational system accidents and testing this framework against real system data.

REFERENCES

Bak, P. 1996. How Nature Works. New York: Springer-Verlag.

Barabosi, A.-L. 2003. Linked. New York: Penguin Books.

Buchanan, M. 2001. Ubiquity. New York: Three Rivers Press.

Grimmett, G. 1999. Percolation. New York: Springer-Verlag.

La Porte, T.R., K.H. Roberts, and G.I. Rochlin. 1987. The self-designing of high reliability organization: aircraft carrier flight operations at sea. Naval War College Review 40(4): 76–90.

Mandelbrot, B. 1982. Fractal Geometry of Nature. New York: W.H. Freeman and Co.

Reason, J. 1997. Managing the Risks of Organizational Accidents. Aldershot, U.K.: Ashgate Publishers.

Stauffer, D., and A. Aharony. 1994. Introduction to Percolation Theory. London and New York: Routledge.

Zipf, G.K. 1949. Human Behavior and the Principle of Least Effort. Cambridge, Mass: Addison-Wesley.

Corporate Cultures as Precursors to Accidents

RON WESTRUM
Department of Sociology
Eastern Michigan University

I strongly believe that precursor situations vary with the corporate culture. In my previous work on cultures and the kinds of accidents they encourage, I have argued that accidents typically have a dominating feature: (1) violations; (2) neglect; (3) overload; or (4) design flaws (Westrum, forthcoming). Although these configurations are admittedly impressionistic, I believe every accident has a dominant character (although combinations certainly exist) (Turner and Pidgeon, 1997). Let me elaborate on these dominant characters a bit more.

Violations are actions that take place in blatant disregard of rules. Of course, rules are often tinkered with to a certain extent, but violations are different. At the 2001 Australian Aviation Psychologists' Symposium Bob Helmreich quoted the remark "checklists are for the lame and the weak." It is a good representation of this attitude. Tony Kern (1999) explores this subject in his book *Darker Shades of Blue*.

Neglect involves the dominance of a known but unfixed problem in an accident configuration. Reason's classic "latent pathogen" probably belongs in this category (Reason, 1990). In this scenario, fixes for problems are ignored, deferred, dismissed, or incomplete. A "dress rehearsal" incident may even take place before an actual accident occurs.

Overload occurs when there is a mismatch between tasks and the resources required to address them. Overload may occur as the result of an organization taking on too large a task. Even though everyone works hard, there is too much work for too few people. Mergers, expansions, downsizing, and reshuffling can all generate overloads. An overload can also occur spontaneously when a team or an individual decides to accept a task that requires more time or skill than they have.

Design flaws occur when conscious decisions lead to unsuspected consequences. Unlike neglect, when a conscious decision is made to ignore a problem, here the problem is unseen, so no one plans for it. Design flaws are insidious, and eliminating all design flaws is very difficult. A design flaw can occur through a failure of imagination or through poor execution of a design.

Any dominant factor can shape an accident for an organization, but I believe their frequency varies systematically with the corporate culture. In my previous work, I have proposed that organizational cultures can be ranged along a spectrum of information flow from pathological to bureaucratic to generative (Westrum, 1994; Turner and Pidgeon, 1997). Organizations with a pathological culture, for instance, have an atmosphere of fear and intimidation, which often reflects intense conflicts or power struggles. The bureaucratic culture, by contrast, is oriented toward following rules and protecting the organization's "turf," or domain of responsibility. Generative organizations are oriented toward high performance and have the most effective information flow.

To me, it follows naturally from the nature of information flow in these cultures that each has particular vulnerabilities (i.e., accident precursors). For instance, a pathological environment encourages overt and covert violations of safety policies. Rogues or "cowboys" pretty much do what they want no matter what the rules are. By contrast, accidents caused by safety violations are rare in a generative environment. Furthermore, generative accidents typically do not show results associated with neglect. Overload is a possibility, but what often catches up generative organizations are design flaws, problems that have been created by conscious decisions whose consequences are not recognized until they have played out in reality. Bureaucratic organizations (most frequently represented in the "systems accident" population) typically fail because they have neglected potential problems or have taken on tasks for which they do not have the resources to do well.

I believe that these tendencies have profound consequences for dealing with accident precursors. The Reason model provides a good general approach to the problem of latent pathogens, but I believe we can do better. One implication of these special vulnerabilities is that even the nature of latent pathogens may differ from one kind of culture to another. By recognizing how cultures differ, we may have a better idea of where to look for problems.

The challenge of a pathological environment is that the culture does not promote safety. In this environment, safety personnel mostly put out fires and make local fixes. The underlying problems are unlikely to be fixed, however. In fact, the pathological organizational environment encourages the creation of new pathogens and rogue behavior. The best techniques in the world can never be enough in an environment where managers practice or encourage unsafe behavior.

In bureaucratic environments, the challenge is a lack of conscious awareness. In the neglect scenario, problems are present, and may even be recognized, but the will to address them is absent. Bureaucratic organizations need to develop a consciousness of common cause, of mutual effort, and of taking prompt action to eliminate latent pathogens. In an overload situation, which has a great potential for failure, the organization needs outside help to cut tasks down to a size it can handle. Groupthink is an ever-present danger in neglect or overload situations, because it can mask problems that need to be faced.

Generative organizations may seem to be accident-proof, but they are not. Generative organizations do not do stupid things, but design flaws are insidious. In the cases of violations, neglect, and overload, the environment provides clear indications to an outside analyst that something is wrong. You can measure the sickness or inefficiency of the culture by tests, observations, and analysis. For instance, there are "symptoms" of groupthink. By contrast, design flaws can be present even when a culture shows no overt symptoms of pathology. Design flaws often come from a failure of what I have called "requisite imagination," an inability to imagine what might go wrong. Even generative cultures suffer from design flaws. In a recent paper, Tony Adamski and I have suggested how requisite imagination can be increased (Adamski and Westrum, 2003). Yet, I believe no system is really capable of predicting all negative consequences. Hence, requisite imagination is more of an art than a science.

We can now look at the differences in confronting latent pathogens in the three different cultural situations. In a generative environment, pointing out (or even discovering) a latent pathogen is usually sufficient to get it fixed. Things are very different in a bureaucratic environment, where inertia or organizational commitments stand in the way of fixing the latent pathogen. When an organization has an overload problem, fixing the problem can be very difficult. In a pathological environment, pointing out a latent pathogen is personally dangerous and may result in the spotter, rather than the pathogen, getting "fixed." I believe that knowing the specific types of failure and their typical generating conditions can help organizations eliminate latent pathogens. If pathogenic situations vary with the environment, then maybe our approach to cleaning them up ought to vary, too.

These observations are impressionistic, but they can be a starting point for further inquiry into links between organizational cultures and the dynamics of latent pathogens. Latent pathogens are constantly generated and constantly removed. Accident reports contain voluminous information about the production of latent pathogens, but we do not know enough about the processes for removing them. The characteristics of healthy environments might be a good topic for a future workshop.

REFERENCES

Adamski A., and R. Westrum. 2003. Requisite Imagination: The Fine Art of Anticipating What Might Go Wrong. Pp. 187–220 in Handbook of Cognitive Task Design, E. Hollnagel, ed. Mahwah, N.J.: Lawrence Erlbaum Associates.

Kern, T. 1999. Darker Shades of Blue: The Rogue Pilot. New York: McGraw-Hill.

Reason, J. 1990. Human Error. Cambridge, U.K.: Cambridge University Press.

Turner, B.A., and N.F. Pidgeon. 1997. Man-Made Disasters, 2nd ed. Oxford, U.K.: Butterworth Heineman.

Westrum, R. 1994. Cultures with Requisite Imagination. Pp. 401–416 in Verification and Validation of Complex Systems: Human Factors, J. Wise et al., eds. Berlin: Springer Verlag.

Westrum, R. Forthcoming. Forms of Bureaucratic Failure. Presented at the Australian Aviation Psychologists Symposium, November 2001, Sydney, Australia.

Appendix B

Workshop Agenda

ACCIDENT PRECURSORS

Linking Risk Assessment with Risk Management

July 17, Lecture Room

8:00 a.m. Continental Breakfast

8:30 a.m. Welcome, Round the Room Introductions and Review of Agenda

Vicki Bier, Professor of Industrial Engineering and Engineering Physics, University of Wisconsin

Howard Kunreuther, Cecilia Yen Koo Professor of Decision Sciences and Public Policy, Wharton School, University of Pennsylvania

James Phimister, J. Herbert Hollomon Fellow, NAE

9:15 a.m. Opening Presentation: The Opportunity of Precursors

James Bagian, Director, National Center for Patient Safety, Veterans Health Administration

9:45 a.m. Q&A

10:00 a.m. Break

10:15 a.m. Session Overview

10:30 a.m. **Session 1: Precursor Detection and Risk Assessment**

Moderator: John Ahearne, Director, Sigma Xi Center

Understanding Accident Precursors

Michal Tamuz, Associate Professor, Center for Health Sciences Research, University of Tennessee Health Science Center

Root-Cause Analysis of Precursors

Bill Corcoran, President, Nuclear Safety Review Concepts Corporation

Nuclear Accident Precursor Assessment: The ASP Program

Martin Sattison, Manager, Risk, Reliability, and Regulatory Support Department, Idaho National Engineering and Environmental Laboratory

12:00 p.m. Breakout Sessions (lunch will be served during breakouts)

1:15 p.m. Breakout Presentations and Plenary Discussion

2:00 p.m. **Session 2: Risk Management and Risk Mitigation**

Moderator: Hal Kaplan, Professor of Clinical Pathology, Columbia University

Designing for Safety

Dennis Hendershot, Senior Technical Fellow, Process Hazard Assessment Department, Rohm & Haas

Human Error and Recovery

Tjerk van der Schaaf, Associate Professor of Human Factors in Risk Control, Eindhoven University of Technology

Harnessing Information and Knowledge Management

John Carroll, Professor of Behavioral and Policy Sciences, Sloan School of Management, Massachusetts Institute of Technology

3:30 p.m. Breakout Session

5:00 p.m. Breakout Presentations

5:45 p.m. Break

6:00 p.m. Reception in the Rotunda

6:30 p.m. Dinner in the Members Room

7:30 p.m. Keynote

Elisabeth Paté-Cornell, Burton J. & Anne M. McMurty Professor and Chair, Department of Management Science and Engineering, Stanford University

8:00 p.m. Adjourn

July 18, Lecture Room

8:00 a.m. Continental Breakfast

8:30 a.m. Chair Comments

9:00 a.m. **Session 3: Linking Risk Assessment with Risk Management**

Moderator: Robert Francis, Senior Policy Advisor, Zucker, Scoutt & Rasenberger

Cross-Industry Applications of a Confidential Reporting Model

Linda Connell, Director, NASA Aviation Safety Reporting System

The Global Aviation Information Network

Chris Hart, Assistant Administrator, Office of System Safety, Federal Aviation Administration

Risk Management and Information Engineering

Yacov Haimes, Professor of Systems and Information Engineering, and Founding Director, Center for Risk Management of Engineering Systems, University of Virginia

10:30 a.m. Break

10:45 a.m. Plenary Discussion

12:00 a.m. Workshop Reflections

Joseph Minarick, Senior Staff Scientist, SAIC

Anita Tucker, Assistant Professor, Operations and Information Management Department, Wharton School, University of Pennsylvania

Deborah Grubbe, Corporate Director, Safety and Health, DuPont

Irv Statler, Project Manager, Aviation Performance Measuring System, NASA

12:40 p.m. Future Directions and Concluding Remarks

Vicki Bier, Professor of Industrial Engineering and Engineering Physics, University of Wisconsin

Howard Kunreuther, Cecilia Yen Koo Professor of Decision Sciences and Public Policy, Wharton School, University of Pennsylvania

1:00 p.m. Lunch

Appendix C

Workshop Participants

John Ahearne
Director, Ethics Program
Sigma-Xi, The Scientific Research Society
99 Alexander Drive
Research Triangle Park, NC 27709

James Bagian
Director
Veterans Affairs National Center for Patient Safety
PO Box 486
Ann Arbor, MI 48106

Dave Balderston
Special Assistant for Risk Management
Federal Aviation Administration
800 Independence Avenue SW
Washington, DC 20591

Patrick Baranowsky
Bureau Chief
Operating Experience Risk Analysis
Nuclear Regulatory Commission
One White Flint North Building
1155 Rockville Pike
Rockville, MD 20852

Vicki Bier
University of Wisconsin
451 Mechanical Engineering Building
1513 University Avenue
Madison, WI 53706

Marilyn Sue Bogner
President and Chief Scientist
Institute for the Study of Human Error, LLC
9322 Friars Road
Bethesda, MD 20817-2308

John Carroll
Professor
School of Management
MIT
77 Massachusetts Avenue
Cambridge, MA 02139

Linda Connell
Director, ASRS/PSRS
NASA Ames Research Center
Moffett Field, CA 94035

Robert Coovert
Nuclear Operations
Exelon Generation Company
4300 Winfield Road, First Floor
Warrenville, IL 60555

William Corcoran
21 Broadleaf Circle
Windsor, CT 06095-1634

Joseph Fragola
Vice President, Principal Scientist
Science Applications International Corporation
265 Sunrise Highway, Suite 22
Rockville Centre, NY 11570

Robert Francis
Senior Policy Advisor
Zucker, Scoutt & Rasenberger
888 17th Street NW
Washington, DC 20006

Jack Fritz
Senior Program Officer
National Academy of Engineering
500 5th Street NW
Washington, DC 20001

Robert Frosch
Senior Research Fellow
Belfer Center for Science and International Affairs
Kennedy School of Government
Harvard University
79 John F. Kennedy Street
Cambridge, MA 02138

Yuri Gawdiak
Program Manager
National Aeronautics and Space Administration
300 E Street SW
Washington, DC 20546

Annetine Gelijns
Associate Professor of Surgery and Health Policy and Management
School of Public Health
Columbia University
600 W. 168th Street, 7th Floor
New York, NY 10032

Deborah Grubbe
Corporate Director, Safety and Health
DuPont Company
1007 Market St., D. 6064
Wilmington, DE 19898

Yacov Haimes
Professor
University of Virginia
PO Box 400736
Olsson Hall, 112a
Charlottesville, VA 22904

Christopher Hart
Assistant Administrator for System Safety
Federal Aviation Administration
800 Independence Avenue SW
Washington, DC 20591

Miriam Heller
Program Director
Division of Civil and Mechanical Systems
National Science Foundation
4201 Wilson Blvd
Arlington, VA 22230

Dennis Hendershot
Senior Technical Fellow
Rohm & Haas
Route 413 and State Roads
P.O. Box 584
Bristol, PA 19007

Sally Kane
Senior Advisor
Directorate for Social, Behavioral, and Economic Sciences
National Science Foundation
4201 Wilson Blvd.
Arlington, VA 22230

Harold Kaplan
Columbia Presbyterian Medical Center
622 W. 168th St., HP4-417
New York, NY 10032

Howard Kunreuther
Professor
The Wharton School
University of Pennsylvania
500 Jon M. Huntsman Hall
Philadelphia, PA 19104

Donald Marksberry
Project Manager
Accident Sequence Precursor Program
Nuclear Regulatory Commission
Mail Stop: T-9C4
Two White Flint North
11545 Rockville Pike
Rockville, MD 20852

Elizabeth Miles
Worldwide Manager, Safety Learning and Development
Johnson & Johnson Corporate Safety
1 J & J Plaza, WH7334
New Brunswick, NJ 08933

Joseph Minarick
Senior Staff Scientist
Science Applications International Corporation
301 Laboratory Road
PO Box 2501
Oak Ridge, TN 37831

Alan Moskowitz
Associate Professor of Clinical Medicine and Health Policy and Management
Columbia University
600 W. 168th St., 7th Floor
New York, NY 10032

Jordan Multer
Human Factors Engineer
Volpe National Transportation Systems Center
55 Broadway
Cambridge, MA 02142

Robert O'Connor
Program Director
Division of Social and Economic Sciences
National Science Foundation
4201 Wilson Blvd.
Arlington, VA 22230

Elisabeth Paté-Cornell
Professor and Chair
Department of Management Science and Engineering
Stanford University
350 Terman Engineering Building
Stanford, CA 94305

James Phimister
J. Herbert Hollomon Fellow
National Academy of Engineering
500 5th Street NW
Washington, DC 20001

D.V. Rao
Los Alamos National Laboratory
PO Box 1663
Los Alamos, NM 87545

Proctor Reid
Associate Director, Program Office
National Academy of Engineering
500 5th Street NW
Washington, DC 20001

Ronald Rardin
Program Director
Division of Design, Manufacture, and Industrial Innovation (ENG/DMII)
National Science Foundation
4201 Wilson Blvd.
Arlington, VA 22230

Claire Reiss
Deputy Executive Director and General Counsel
Public Entity Risk Institute
11350 Random Hills Road, Suite 210
Fairfax, VA 22030

Irving Rosenthal
Member
Chemical Safety and Hazard Investigation Board
2175 K Street, Suite 400
Washington, DC 20037

Martin Sattison
Idaho National Engineering and Environmental Laboratory
IF EROB 341
MS 3870
Idaho Falls, ID 83415

Herbert Schlickenmaier
National Aeronautics and Space Administration
300 E Street SW
Washington, DC 20546

Jeffrey Shackelford
Staff Member
Defense Nuclear Facilities Safety Board
625 Indiana Avenue NW
Washington, DC 20004

Irving Statler
Aviation System Monitoring and Modeling
NASA Ames Research Center
Moffett Field, CA 94035

Margaret Sweeney
Bureau of Transportation Statistics
400 7th Street SW, Room 3103
Washington, DC 20590

Michal Tamuz
Center for Health Services Research
66 North Pauline, Suite 463
Health Science Center
University of Tennessee
Memphis, TN 38163

Phyllis Thompson
Program Analyst
Chemical Safety Board
2175 K Street NW, Suite 400
Washington, DC 20037

Anita Tucker
Assistant Professor
The Wharton School
University of Pennsylvania
551 JMHH
3730 Walnut Street
Philadelphia, PA 19104-6340

Tjerk van der Schaaf
Professor
Eindhoven University of Technology
Den Dolech 2
P.O. Box 513
5600 MB Eindhoven,
The Netherlands

William Vesely
Technical Risk Manager
National Aeronautics and Space Administration
300 E Street SW, Code QE
Washington, DC 20024

Lawrence Yuspeh
Director of Research and Development
Louisiana Workers' Compensation Corporation
2237 South Acadian Thruway
Baton Rouge, LA 70808

Rae Zimmerman
Professor
New York University
Robert F. Wagner Graduate School of Public Service
411 Lafayette St, 3rd Floor
New York, NY 10003

Appendix D

A Note on Definitions

The committee recognizes that terminology used by parallel disciplines can have slightly different meanings and interpretations. This can make it difficult to conduct a cross-disciplinary dialogue, because a sentence may be interpreted one way when the speaker means something subtly, yet distinctly, different.

We ask that participants be aware of and sensitive to this challenge. When speaking, recognize that the audience may not be familiar with your area of expertise, and when possible, introduce your point by discussing it in the context of your field and experience. Similarly, when listening, pay special attention to terminology and the context in which words are used. If the usage of a word differs from what you expect, even slightly, please feel free to ask for clarification; such clarifications often lead to greater insight for everyone.

An obvious example of terminology that can cause confusion is the phrase *accident precursors*, or even the word *precursors*. In our discussions with committee members, panelists, and participants, we found that there are many different interpretations of these terms. Some people define accident precursors as the events that immediately precede and lead up to an accident; others include organizational and cultural shifts in the definition; and some also include macro-events outside an organization in the definition.

In the proposal for this project, we defined precursors as events in the accident chain. We found that this narrow definition aided our discussions; for instance, we could discuss how an organization could look for precursors without the organization itself being considered a potential precursor. For this workshop, we encourage you to use this definition and to explain alternative usages when necessary.

Near miss and its analogs, *near hit* and *close call*, are other terms that are likely to arise frequently during workshop discussions. Although near misses are clearly related to precursors, we have tried to distinguish them from precursors, and we encourage you not to use them interchangeably. One way to define a near miss (or, equivalently, a near hit or close call) is as an almost complete progression of events—a progression that, if one other event had occurred, would have resulted in an accident. A near miss might consist of one or more precursors that did occur, and one that did not. A near miss can be considered a particularly severe precursor. Understanding the subtle distinction between a near miss and a precursor can facilitate workshop discussions.

In the papers in the agenda book, Michal Tamuz and Bill Corcoran address some of the challenges to defining and understanding precursors. The paper by Petrie, "Do You See What?: The Epistemology of Interdisciplinary Inquiry," included in the back of the agenda book, suggests other ways to facilitate interdisciplinary discussions.

Appendix E

Biographies of Authors

James P. Bagian has extensive experience in the fields of human factors in aviation and patient safety. He was chosen as the first director of the Department of Veterans Affairs' National Center for Patient Safety in 1998. A former astronaut, Dr. Bagian flew on two Space Shuttle missions and supervised the recovery of *Challenger* following the explosion in 1986. He also led the development of the Space Shuttle Escape System now in use. Dr. Bagian holds a B.S. in mechanical engineering from Drexel University and an M.D. from Thomas Jefferson University. He is a member of the National Academy of Engineering and is on the faculty of the Uniformed Services University of Health Sciences and University of Texas Medical Branch. Dr. Bagian received the American Medical Association 2001 Dr. Nathan S. Davis Award for outstanding public service in the advancement of public health and the Association of American Medical Colleges first annual Innovations Award in 2001. In 2002, he received the Frank Brown Berry Prize in Federal Healthcare, which recognizes a military or federal physician who has made a significant contribution to health care in the United States.

John S. Carroll is professor of behavioral and policy sciences at the Sloan School of Management, Massachusetts Institute of Technology (MIT). He received a B.S. in physics from MIT and a Ph.D. in social psychology from Harvard. Prior to joining the faculty of the Sloan School in 1983, he taught in the psychology departments of Carnegie-Mellon University and Loyola University of Chicago and was a visiting associate professor at the Graduate School of Business, University of Chicago. Professor Carroll has published four books and numerous articles in several areas of social and organizational psychology. Much of his research has been focused on individual and group decision making, the relationship between cognition and behavior in organizational contexts, and the processes that link individual, group, and organizational learning. Professor Carroll is a fellow of the American Psychological Society.

Linda Connell is the director of the National Aeronautics and Space Administration (NASA) Aviation Safety Reporting System, program manager/director of the NASA/Veterans Affairs Patient Safety Reporting System, and, since 1981, a research psychologist for NASA Ames Research Center. Ms. Connell has participated in numerous studies with domestic and international research teams exploring human-factors issues in aviation and other environments. She completed her bachelor's degree in nursing at the University of Colorado and her master's degree in experimental psychology at San Jose State University. Her graduate thesis on psychophysiological countermeasures to jet lag was completed at the NASA Ames Human Research Facility.

William R. Corcoran is president of the Nuclear Safety Review Concepts Corporation in Windsor, Connecticut, founded in 1993. The organization's motto is "Saving lives, pain, assets, and careers through thoughtful inquiry." His company provides root-cause training, mentoring, advice, leadership, and assistance to a variety of high-hazard industries, including nuclear power, natural gas, electricity transmission/distribution, and fossil-fuel power generation. Dr. Corcoran is author of *The Phoenix Handbook* (Nuclear Safety Review Concepts Corporation, 1998), a comprehensive guide to root-cause analysis, and editor of "The Firebird Forum," an electronic newsletter on root-cause organizational learning. He has served on more than a dozen off-site safety review committees for nuclear plants and chaired the American Nuclear Society Nuclear Reactor Safety Division. His work on critical safety functions has influenced the industry's emergency procedures.

Robert A. Frosch earned a Ph.D. in physics from Columbia University in 1952. From 1951 to 1963, as director of Hudson Laboratories at Columbia, he did research on ocean acoustics. From 1963 to 1966, he was director for nuclear test detection, then deputy director of the U.S. Department of Defense Advanced Research Projects Agency. From 1966 to 1973, Dr. Frosch was assistant secretary of the Navy (Research and Development). In 1973, he became assistant executive director for the United Nations Environment Programme, and in 1975, he became associate director for applied oceanography at the Woods Hole Oceanographic Institution. From 1977 to 1981, he was administrator of National Aeronautics and Space Administration. In 1981, he became president of the American Association of Engineering Societies, and in 1982, he was vice president of General Motors Corporation (GM) in charge of research laboratories. Dr. Frosch retired from GM in 1993 and joined the John F. Kennedy School of Government of Harvard University. He is a member of Phi Beta Kappa and Sigma Xi, the National Academy of Engineering, and the American Academy of Arts and Sciences and a Foreign Member of the Royal Academy of Engineering.

Yacov Y. Haimes, professor of systems and information engineering at the University of Virginia, is founding director (1987) of the University of Virginia Center for Risk Management of Engineering Systems and Lawrence R. Quarles Professor in the School of Engineering and Applied Science. As a member of the faculty at Case Western Reserve University for 17 years, he was chair of the Systems Engineering Department. During a sabbatical year in 1977–1978, as a AAAS/AGU Congressional Science Fellow, he joined the staff of the Executive Office of President Carter, and later the staff of the House Science and Technology Committee. Dr. Haimes is the recipient of several major awards in his field, including the Distinguished Achievement Award from the Society for Risk Analysis; the Georg Cantor Award from the International Society on Multiple Criteria Decision Making; the Norbert Weiner Award from IEEE Systems, Man and Cybernetics; and the Warren A. Hall Medal from the Universities Council on Water Resources. He is also a fellow of many professional societies and was president of the Society for Risk Analysis. Dr. Haimes has published more than 200 articles and technical papers and is the author/coauthor of six books and editor of 20 volumes. His most recent book is *Risk Modeling, Assessment, and Management* (John Wiley & Sons, 1998, 2nd ed. 2004).

Christopher A. Hart became the assistant administrator for the Federal Aviation Administration (FAA) Office of System Safety in 1995. Reporting directly to the FAA administrator, the Office of System Safety provides data, analytical tools and processes, safety risk assessments, and other assistance to support numerous FAA and worldwide aviation safety programs; spearheads industry-wide safety activities, such as the Global Aviation Information Network (GAIN); and helps identify key safety issues and emerging trends that affect aviation safety. Mr. Hart has a law degree from Harvard Law School and a master's degree (magna cum laude) in aerospace engineering from Princeton University. He is a member of the District of Columbia Bar Association and the Lawyer-Pilots Bar Association and a pilot with commercial, multi-engine, and instrument ratings.

Dennis C. Hendershot is a senior technical fellow in the Process Hazard Assessment Department of the Rohm and Haas Company Engineering Division in Croydon, Pennsylvania. He has been involved with the development and application of hazard analysis, risk management, and safety engineering tools, with particular emphasis on inherently safer design, process hazard analysis, and quantitative risk analysis. He received a B.S. in chemical engineering from Lehigh

University and an M.S. in chemical engineering from the University of Pennsylvania. Mr. Hendeshot is a fellow of the American Institute of Chemical Engineers (AIChE) and is currently on the AIChE Board of Directors. He is a past chair of the AIChE Safety and Health Division and the AIChE Loss Prevention Programming Committee. He serves on the editorial review boards of *Process Safety Progress*, *Chemical Engineering Progress*, and *Journal of Loss Prevention in the Process Industries*. He has been active in the Center for Chemical Process Safety (CCPS), as a member and chair of the Risk Assessment Subcommittee, chair of the Hazard Evaluation Procedures Subcommittee, a member of the Inherently Safer Process Subcommittee, member and chair of the Undergraduate Education Subcommittee, and a member of the CCPS Managing Board. Mr. Hendershot is a member of the Division of Chemical Health and Safety and the Division of Environmental Chemistry of the American Chemical Society.

Lisette Kanse is a postdoctoral research fellow at the Key Centre for Human Factors and Applied Cognitive Psychology at the University of Queensland in Brisbane, Australia. She has just completed her Ph.D. at the Eindhoven University of Technology, focusing on the processes involved in recovering from failures. She has also worked as a consultant in safety management and human factors at several chemical process plants and in the rail sector. Dr. Kanse's research interests include safety management, failure recovery, human factors, incident reporting and investigation, and organizational learning.

Elisabeth Paté-Cornell is the Burt and Deedee McMurtry Professor in the School of Engineering and has been chair of the Department of Management Science and Engineering at Stanford University since its creation in January 2000. From 1978 to 1981, she was assistant professor of civil engineering at Massachusetts Institute of Technology, and from 1981 to 1999, she was a faculty member at Stanford in the Department of Industrial Engineering and Engineering Management. Her primary areas of teaching and research are engineering risk analysis and risk management, decision analysis, and engineering economy. Her recent research has focused on the inclusion of organizational factors in probabilistic risk analysis models with application to the management of the protective tiles on the space shuttle, offshore platforms, and anesthesia during surgery. She is currently working on the trade-offs between management and technical failure risks with application to the design of space systems, and on probabilistic methods of threat analysis. Dr. Pate-Cornell is a member of the National Academy of Engineering (NAE) and NAE Council, the President's Foreign Intelligence Advisory Board, the Air Force Science Advisory Board, and the California Council on Science and Technology. She is a past president and fellow of the Society for Risk Analysis. She is currently an elected member of the Stanford University Senate.

Martin B. Sattison is manager of the Risk, Reliability, and Regulatory Support Department at the Idaho National Engineering and Environmental Laboratory (INEEL). He has been associated with the Nuclear Regulatory Commission's Accident Sequence Precursor (ASP) Program for 10 years as the lead analyst, project manager, and program manager for the development of the Standardized Plant Analysis Risk Models currently used for the ASP Program. Dr. Sattison has had 25 years of experience in the nuclear field and 20 years of experience in probabilistic risk assessment (PRA). He was a member of the International Space Station PRA Peer Review Committee and is currently a consultant to the National Aeronautics and Space Administration on the Space Shuttle PRA. In addition, he recently assisted with the *Columbia* accident investigation.

Michal Tamuz is an organizational sociologist with research interests in risk management, improving patient safety, organizational learning, and decision making. She developed an interest in how organizations cope with uncertainty when she lived in Israel and is currently studying how hospitals and hospital pharmacies learn from medication errors, as part of a study at the University of Texas on translating safety practices from aviation to health care. Her research focuses on near-accident reporting in an array of high-hazard industries, such as aviation, chemical manufacturing, nuclear power, and blood banks. In her research, she explores how organizations learn under conditions of ambiguity and scarcity of experience. Dr. Tamuz received a Ph.D. in sociology from Stanford University. She is a faculty member at the Center for Health Services Research and an associate professor in the Department of Preventive Medicine, College of Medicine, University of Tennessee Health Science Center, Memphis, Tennessee.

Ron Westrum is professor of sociology and interdisciplinary technology at Eastern Michigan University. A graduate of Harvard and the University of Chicago, he is the author of numerous articles and three books, the most recent on the culture of the China Lake Naval Weapons Laboratory. Dr. Westrum, who specializes in the study of corporate cultures relevant to system safety, is a frequent speaker on systems safety in the aviation, medical, and nuclear industries and has been a consultant to many large organizations, including General Motors and Lockheed Martin.

Tjerk van der Schaaf has been an associate professor of human factors in risk control, Faculty Technology Management, Department of Human Performance Management, Eindhoven University of Technology, for the past seven years. He has also been coordinator, Eindhoven Safety Management Group; assistant professor of human error and industrial safety; and researcher in the Psychology Department, TNO-Institute of Human Factors, Soesterberg, The Netherlands. Dr. van der Schaaf has a Ph.D. from Eindhoven University of Technology.

Appendix F

Biographies of Committee Members

Vicki Bier (co-chair) is a leading researcher on accident precursors. She was editor and organizer of *Accident Sequence Precursors: Risk Assessment and Probabilistic Risk Analysis* (University of Maryland, 1998) and has published extensively on modeling precursors and precursor analysis. Dr. Bier is a professor of industrial engineering and engineering physics at the University of Wisconsin-Madison.

Howard Kunreuther (co-chair) is the Cecilia Yen Koo Professor of Decision Sciences and Public Policy at the Wharton School, University of Pennsylvania, as well as codirector of the Wharton Risk Management and Decision Processes Center. He is currently on sabbatical at Columbia University as a visiting research scientist. Dr. Kunreuther has a long-standing interest in the management of low-probability high-consequence events related to technological and natural hazards and has published extensively on the topic. He was a member of the National Research Council Board on Natural Disasters and chaired the H. John Heinz III Center Panel on Risk, Vulnerability and True Costs of Coastal Hazards and is a recipient of the Elizur Wright Award for the most significant publication in the literature of insurance. Dr. Kunreuther is a distinguished fellow of the Society for Risk Analysis (SRA) and was awarded the SRA Distinguished Achievement Award in 2001. He is the author, with Paul Freeman, of *Managing Environmental Risk through Insurance* (Kluwer Academic Publishers, 1997), coeditor (with Richard Roth, Sr.) of *Paying the Price: The Status and Role of Insurance Against Natural Disasters in the United States* (National Academy Press, 1998), and coeditor (with Steve Hoch) of *Wharton on Making Decisions* (John Wiley & Sons, 2001).

John F. Ahearne is adjunct professor of civil and environmental engineering and lecturer on public policy studies at Duke University. He is also director of the Sigma Xi Center (The Scientific Research Society). Dr. Ahearne earned his Ph.D. in physics from Princeton University and is an expert on nuclear power and nuclear weapons. From 1978 to 1983, he was a commissioner of the U.S. Nuclear Regulatory Commission and was chairman from 1979 to 1981. In 1997–1998, he was deputy assistant secretary of energy in the White House Energy Office, and from 1972 to 1977, he was deputy and principal deputy assistant secretary of defense working on weapons systems analysis. He served in the U.S. Air Force from 1959 to 1970, and has worked at the U.S. Air Force Weapons Center on nuclear weapons effects and taught at the U.S. Air Force Academy, Colorado College, and the University of Colorado. Dr. Ahearne has served on and chaired numerous government research and policy making committees concerning nuclear energy, nuclear weapons, and the disposal of nuclear waste.

Robert Francis, who was appointed vice chairman of the National Transportation Safety Board (NTSB) by President Clinton in 1995, was involved in many transportation accident investigations; Mr. Francis also chaired a number of NTSB public hearings. In addition to his investigative work and other NTSB duties, he has been an active member of the Air Transport Association of America Steering Committee on Flight Operations Quality Assurance Programs and the Flight Safety Foundation ICARUS Committee, a group of aviation experts from around the world who gather to share ideas on reducing human error in the cockpit. Prior to his appointment to the NTSB, Mr. Francis was senior representative of the Federal Aviation Administration (FAA) in Western Europe and North Africa. As representative of the FAA administrator, he worked extensively on aviation safety and security issues with U.S. and foreign air carriers, transportation governmental authorities, aircraft manufacturers, and airports. Mr. Francis received the Aviation Week and Space Technology 1996 Laurels Award and was recognized by both the U.S. Navy and U.S. Coast Guard for meritorious service in the TWA Flight 800 investigation. He received his A.B. from Williams College, attended Boston University and the University of Ibadan, Nigeria, and holds a commercial pilot certificate with instrument and twin-engine ratings. He is a member of the French Academy of Air and Space, a fellow of the Royal Aeronautical Society, a member of the Wings Club of New York, a trustee of the Aero Club of Washington, a member of the Board of Directors for Women in Aviation, International, and the Executive Council of NASA's Aviation Safety Program. In 2000, Mr. Francis left the NTST and joined the Aerospace Safety Advisory Panel as a consultant.

Harold S. Kaplan is professor of clinical pathology at the College of Physicians and Surgeons, Columbia University, and director of transfusion medicine, New York Presbyterian Hospital, Columbia-Presbyterian Medical Center. A graduate of Oberlin College, he earned his M.D. from Albert Einstein College of Medicine. He completed his postgraduate training in pathology at the Columbia-Presbyterian Medical Center. Dr. Kaplan is the principal investigator for the NHLBI RO1 Grant for the development and implementation of an event reporting system (MERS-TM) to improve the safety of the blood supply. His current research is focused on establishing the usefulness of standardized medical event reporting for error prevention and management.

Harry McDonald is Distinguished Professor, Chair of Excellence in Engineering, University of Tennessee at Chattanooga, where he conducts scholarly research and provides advice and assistance to the university faculty and students. From 1996 to 2002, as center director, NASA Ames Research Center, he was responsible for defining and overseeing the NASA Ames Center of Excellence for Information Technologies, including all research programs, approximately 1,500 civil servants, 3,000 contractor employees, and an annual budget approaching $1 billion. The center is heavily involved in supercomputing, information technologies, and aerospace and space-science research. Dr. McDonald received a D.Sc. in engineering and a B.Sc. in aeronautical engineering, with Honors, both from the University of Glasgow.

Elizabeth Miles is worldwide manager of safety learning and development for Johnson & Johnson (J&J). Her responsibilities include: the development of safety learning and development strategies for the J&J family of companies; core competency for behavior-based safety; and the creation of the incident investigation process, "Learning to Look." Ms. Miles has been with J&J in various capacities since 1984. She has an M.S. in organizational dynamics from the University of Pennsylvania, an M.A. in the history of ideas from Johns Hopkins University, and a B.S.in biology from the University of Maryland.

Elisabeth Paté-Cornell is the Burt and Deedee McMurtry Professor in the School of Engineering and has been chair of the Department of Management Science and Engineering at Stanford University since its creation in January 2000. From 1978 to 1981, she was assistant professor of civil engineering at Massachusetts Institute of Technology, and from 1981 to 1999, she was a faculty member at Stanford in the Department of Industrial Engineering and Engineering Management. Her primary areas of teaching and research are engineering risk analysis and risk management, decision analysis, and engineering economy. Her research in recent years has focused on the inclusion of organizational factors in probabilistic risk analysis models with application to the management of the protective tiles on the space shuttle, offshore energy platforms, and anesthesia during surgery. She is currently working on trade-offs between management and technical failure risks in the design of space systems and on probabilistic methods of threat analysis. Dr. Paté-Cornell is a member of the

National Academy of Engineering (NAE) and the NAE Council, the President's Foreign Intelligence Advisory Board, the Air Force Science Advisory Board, and the California Council on Science and Technology. She is a past president and fellow of the Society for Risk Analysis and is currently an elected member of the Stanford University Senate. She received her degree in computer science from the Institut Polytechnique, Grenoble, France, and a master's degree in operations research and a doctorate in engineering-economic systems, both from Stanford University.